AF842515

Auckland Unplugged

Auckland Unplugged

Coping with Critical Infrastructure Failure

Lindy Newlove,
Eric Stern,
and Lina Svedin

LEXINGTON BOOKS
Lanham • Boulder • New York • Oxford

LEXINGTON BOOKS

Published in the United States of America
by Lexington Books
A Member of the Rowman & Littlefield Publishing Group
4501 Forbes Boulevard, Suite 200, Lanham, Maryland 20706

PO Box 317
Oxford
OX2 9RU, UK

Copyright © 2003 by Lexington Books

All rights reserved. No part of this publication may be reproduced, stored in a retrieval system, or transmitted in any form or by any means, electronic, mechanical, photocopying, recording, or otherwise, without the prior permission of the publisher.

British Library Cataloguing in Publication Information Available

Library of Congress Cataloging-in-Publication Data

Newlove, Lindy, 1967-
Auckland unplugged : coping with critical infrastructure failure /
Lindy Newlove, Eric Stern, and Lina Svedin.
p. cm.
Includes bibliographical references and index.
ISBN 0-7391-0486-1 (alk. paper)
1. Emergency management—New Zealand—Auckland. 2. Electric power failures—New Zealand—Auckland. 3. Electric power distribution—New Zealand—Auckland—Evaluation. I. Stern, Eric, 1965- II. Svedin, Lina, 1974- III. Title.
HV551.5.N45N49 2003
333.793'2'099324—dc21 2002155645

Printed in the United States of America

∞™ The paper used in this publication meets the minimum requirements of American National Standard for Information Sciences—Permanence of Paper for Printed Library Materials, ANSI/NISO Z39.48-1992.

Contents

Preface

As we place the final, affectionate, brush strokes on the canvas that is this study, we gratefully remember and acknowledge the generous support of the many friends and colleagues in Sweden, Holland, and New Zealand who have in one way or another helped us to get here. The idea for the study came from Jan Lundberg of the Swedish Agency for Civil Emergency Planning (ÖCB), which generously provided funding for the project. Jan and his colleagues at ÖCB also provided us with some initial materials and introduced us to the civil defense community in New Zealand. Mats Kero and Jan Wisén of the Stockholm Fire Department kindly briefed us on their experiences as 'observers' during the power crisis and shared their Auckland contacts with us. Staffan Molin and Georg Fischer of the Swedish National Defence Research Agency (FOI) shared the fruits of their internet data-mining with us and provided technical counterpoint to our more social scientific approach to crisis analysis. Veiko Kekki and Håkan Stomsjö of the Swedish National Grid Company (Svenska Kraftnät AB) read our report and provided us with feedback.

We are most grateful for the warm Kiwi hospitality we received in New Zealand. We appreciate the efforts made by our interviewees (see the bibliography of the study for a complete list) and other informants in New Zealand. General Director John Norton of the Ministry of Civil Defence in Wellington generously sponsored our participation in the annual Civil Defence conference at Lake Taupo, which coincided with our visit. The Auckland office of the Ministry kindly allowed us to use their facilities as a base of operations and allowed us liberal access to their archives. Special thanks to Frank Arnott, John Titmus and Kelvin McMinn who went out of their way to deal with various logistical problems. Finally, many, many thanks to our good friend Fred Henare of the Auckland Ministry of Civil Defence who acted as our guide to Auckland and provided us with many insights into the power crisis and life in New Zealand.

Paul 't Hart, Jos Rijpma, Werner Overdijk and Menno van Duin of the Crisis Research Center in Leiden/The Hague read the manuscript and gave us useful, critical comments. Our colleagues at CRiSMART provided similar services, special thanks going to Jesper Grönvall and Daniel Nohrstedt who went over parts of the manuscript with a fine-toothed comb. Stephanie Young helped with language and copy editing.

We also wish to thank the steering group of the Crisis Management Baltic/Europe research program for moral and administrative support, especially Harald Törner and Bo Richard Lundgren. Finally, we wish to thank Bengt Sundelius for suspending his disbelief when we told him that New Zealand was, for all practical purposes, part of the Baltic Sea Area.

—Stockholm, August 2002.

Abbreviations

Auckland Electric Power Board	AEPB
Auckland Engineering Lifelines Project	AELP
Auckland Energy Consumer Trust	AECT
Auckland Energy Limited	AEL
Auckland Regional Council	ARC
Auckland Regional Services Trust	ARST
Central Business District	CBD
Domestic and External Security Co-ordinator	DESC
Electricity Corporation of New Zealand	ECNZ
Electricity Industry Reform Act	EIR Act
Energy Information Administration	EIA
Electricity Supply Authority	ESA
Emergency Management Group	EMG
Energy Efficiency and Conservation Authority	EECA
External Relations Defence and Security Committee	ERDSC
Infrastructure Auckland	IA
Karangahape Road Business Association	K'Rd. Business Assn.
Local Government Amendment Act	LGAA
Local Government New Zealand	LGNZ
Maria Governance Board	MGB
Marketplace Company	M-co.
Member of Parliament	MP
Metering and Reconciliation Information Agreement	MARIA
Ministry of Civil Defence	MoCD

Mixed Member Proportional	MMP
Municipal Electricity Department	MED
New Zealand Electricity Division	NZED
New Zealand Electricity Market	NZEM
New Zealand Television	NZTV
Normal Accident Theory	NAT
Rand Merchant Bank of Australia	RMBA
Situation Reports	SITREPS
State-Owned Enterprise	SOE
Statement of Corporate Intent	SCI
Symbolic-Political	SP
Technical-Managerial	TM
Territorial Local Authority	TLA

1

Introduction

1.1 THE AUCKLAND BLACKOUT

Electricity is the lifeblood of the modern city. Virtually all of the socio-technical systems which maintain public order, quality (and even support) of life—and commerce—are dependent upon a reliable supply of electricity. Computers, climate-control systems, refrigeration of food, transportation, sanitation, tele- and mass communication—all are liable to disruption when electrical power fails. Our fear of darkness and chaos striking at the heart of our cities has been fuelled by Hollywood and burned into our retinas by film footage of plunderers on a rampage like the one during the New York City blackout of 1977. For city governments, business communities, and citizens alike, a prolonged black- or brownout is a nightmare scenario.

In early 1998, the city of Auckland, New Zealand, suffered a prolonged and profound disruption of its electricity supply. The central business district (CBD) of Auckland, the governmental hub of the city and the commercial/financial heart of the country was deprived of reliable access to electricity for over two months. This infrastructural failure rapidly escalated into a major crisis for the city government, specialized agencies, corporate actors (not least Mercury Energy, the stricken power provider), small businesses, and citizens alike.

The ten-week power outage took place in the Auckland CBD between the end of February and the middle of April, 1998 (MoCD, 1998). Thus the prolonged incident took place during high summer in New Zealand, in contrast to other comparable incidents which have taken place under winter conditions such as the Canadian ice storm of 1998 (Newlove, 1998) and the Atlanta ice storm of 2000. While the warm weather conditions preempted certain typical difficulties—such as heating homes and businesses—the outage took place during a heat wave and affected a zone containing many high-rise buildings heavily dependent upon electricity for air conditioning, elevators, lighting, and sanitation.

The power failure occurred when, over the span of two weeks, the four main power cables feeding power into the district collapsed. The area affected, the CBD, is a 3.5-square-kilometer area lively with intensive business and commerce during the day and an active entertainment scene attracting nightly up to an estimated seven thousand people from other parts of Auckland (Davis, 1998a). The area has a comparatively low number of permanent residents to other parts of the city but is however, a part of town where several of the largest hotels are located. At the time of the power outage some four thousand residents and sixty-three thousand employees in ten thousand businesses were calculated to have been included in the area (MoCD, 1998). Though no formal declaration of emergency was issued during the power outage, the situation was regarded as extremely serious, exceptional resources were allocated to crisis management, and the question was given top priority by a wide range of local actors from the mayor of the city on down. Although there were no reported deaths as a result of the power outage, this positive outcome of the crisis was partly the result of a vigilant and proactive crisis response on the part of a number of the responsible agencies (MoCD, 1998).

Arising in the wake of a drastic and highly politicized restructuring of the energy sector and in the midst of a civil defence reform process, the crisis posed a panorama of threats and opportunities to actors and stakeholders alike. The four Ps (public health, public order, public safety, and prosperity) were seen to be seriously threatened in the crisis as was public confidence in the power provider and potentially, in the city government. Persistent uncertainties about the duration of the crisis and the viability of possible technical solutions exacerbated the urgent and ongoing problem of calibrating appropriate crisis responses at various stages of the power crisis. The technical complexity of the issues created a dependency on expert assessments, which were largely provided by Mercury Energy itself. The dynamics of the case placed heavy demands on actors in terms of crisis coordination and communication—providing examples of both best, and less good, practices of crisis decision making and communication. The Auckland blackout of 1998 represents a significant opportunity for studying and learning about infrastructural failures and crisis management in an urban setting—topics which are increasingly emphasized in the crisis management literature (cf. Rosenthal et al., 1994; Rosenthal, Boin, and Comfort, 2001; Castenfors and Svedin, 2001).

The case provides insights into a number of central challenges of crisis governance in post-industrial, democratic societies:

- Finding an appropriate division of responsibility and labor between public and private sector actors when critical infrastructure unexpectedly collapses
- Crafting and coordinating a crisis response which addresses perceived threats to community values and avoids the twin perils of under-reaction (e.g "passivity" or "paralysis") and overreaction (e.g., "crying wolf" or "political grandstanding")
- Coping with competence/authority discrepancies under stress—those who have expert knowledge of the technical issues rarely have the authority to make policy; those who have the authority generally lack the technical expertise to comprehend the subtleties and uncertainties of the issues at stake
- Maintaining credibility and legitimacy when facing acute, ill-structured, problems in politicized, "mass-mediated," and highly uncertain environments

These challenges are by no means specific to Auckland or to the problem of coping with urban blackouts. Rather they are general and recurring problems faced by crisis managers around the world.

1.2 STUDYING INFRASTRUCTURAL FAILURES: A CRISIS MANAGEMENT APPROACH

As we have already noted above, critical infrastructure failures (such as the Auckland power crisis) have profound implications for citizens and for those who govern in their name. When such lifelines are in jeopardy, there is an immediate threat to the well-being and prosperity of society and the legitimacy of the public administration and its political masters. That being the case, it is not surprising that social scientists from a number of disciplines have examined the phenomenon in the hopes of improving our knowledge of why critical systems fail, and how such systems can be made more reliable. For example, Barry Turner and Nick Pidgeon (1997) have looked at the anthropogenic factors allow socio-technical disasters to incubate and escalate (cf. Maurino et al, 1995; Snook, 2000). Charles Perrow's (1999) work on normal accidents directs our attention to the vulnerabilities inherant in the highly complex and tightly coupled large scale social systems which are an integral part of turn of the millennium society (cf. Sagan, 1993). Todd La Porte, Gene Rochlin, Karl Weick, and others associated with the so-called high-reliability school have sought to identify the organizational culture and practices associated with the successful operation of high risk enterprises over long periods of time, seeking exemplars that promise to help others conduct such operations more safely (LaPorte and Consolini, 1991). Other researchers such as the late Aaron Wildavsky (1988) emphasize the prohibitive costs and virtual impossiblity of eliminating the risk of critical infrastructure (and high-risk industrial) failure and suggest that society must cultivate resilience—the capacity to withstand and recover from sociotechnical disasters (Dynes and Tierney, 1994; Quarantelli, 1998).

A particularly interesting strand of the interdisciplinary literature focuses on the challenge of making and implementing policy under the adverse conditions associated with disasters and crises (Dror, 1986; Comfort, 1988; Rosenthal, 't Hart and Charles, 1989). In fact, in this literature, a crisis is often defined in terms of the subjective predicament experienced by those responsible for coping with a given contingency, which could be a natural disaster, an acute economic problem, terrorism, an escalating civil or international conflict, or (as in the case of the Auckland power outage) a critical infrastructure failure. These contingencies—while different in many respects—pose similar challenges of decision making and communication to those who act in the name of the state. Actors at various levels of national and regional administration are likely to perceive these incidents as characterized by *urgency*, *threats to core values*, and *uncertainty*—in other words as *crises* to which they must respond (cf. Hermann, 1963, Brecher, 1993; Rosenthal, Charles and 't Hart, 1989; Stern, 1999). For those caught in the "hot seat" (Flin, 1996) in such situations, political and bureaucratic survival may well be on the line as the crisis unfolds.

In the realm of research, as in the realm of practice, crisis management is a term which is used to refer to many different things and thus requires some specification. For example, in the literature on international relations, the term *crisis management* in often used to describe the efforts by parties to an escalating international conflict to protect vital interests while seeking to prevent an uncontrolled escalation to war (cf. Allison, 1971; Lebow, 1998; George, 1991). In the literature on corporate crises, crisis management often focuses on the problem of communicating with stakeholders and consumers in a manner which protects the image of the threatened organization (cf. Mittroff, Pearson, and Harrington, 1996; Regester and Larkin, 1998).

For our present purposes, crisis management may be seen as a broad umbrella concept which refers to a number of activities relevant to cultivating societal capacity to cope with crises in general and critical infrastructure failures in particular. These include:

- planning and organizational preparedness
- crisis prevention/mitigation
- crisis decision making (in the acute phase)
- crisis communication
- post-crisis accountability and learning

While there are a number of ways of going about studying these various aspects of crisis management, we will apply the process-tracing methodology developed in Sundelius, Stern, and Bynander (1997) and refined in the OCB-sponsored research program Crisis Management Baltic/Europe: National Crisis Management in an International Perspective (e.g., Stern and Bynander, 1998; Stern, 1999) as a means of organizing our inquiry into the Auckland power crisis.

This methodology entails:

a) a detailed reconstruction of the crisis events drawing upon available government documents and reports, mass media sources (electronic and print), and interviews with involved decision makers, stakeholders, and citizens;[1]
b) a dissection of the case into a series of crucial occasions for decision—a series of pressing "what do we do now?" problems which arise during the period of the crisis, challenge the coping capacity of decision makers, and have the potential to shift the course of the events for the better or for the worse;
c) thematic analyses of key phenomena such as leadership, coordination, crisis communication, and mass media relations, etc. (see the sections below on "thematic analysis" and "structure of the study" for a complete list of themes addressed in this study); and
d) comparing and contrasting case findings with others documented in the Crisis Management Europe case bank and the international literature. In addition, several themes that have not been particularly emphasised in the Crisis Management Baltic/Europe research program are highlighted:
   - interaction among public, semi-public, and private sector actors and stakeholders in crisis management;
   - the "victim" perspective—crisis coping viewed from the vantage point of the individual citizen and or small business enterprise;
   - the process of crisis development—why did it prove impossible to prevent the initial cable failures from developing into a full-blown and prolonged power crisis?

1.3 URBAN CRISIS MANAGEMENT

Within the cross-disciplinary field of urban studies, this study on urban crisis management is in various ways linked to the interests and issues identified in urban planning, city management, cultural geography, and urban sociology.

1.3.1 Urban Planning

A great deal of the literature on urban policy and urban development belongs in the area of urban planning in general. This literature focuses on urban policies and programs concerning issues of geographical and economic development, and environmental preservation in urban areas (North, Valler, and Wood, 2001). The literature in these areas has flourished in the United States during the latter half of the last century and has gained a strong foothold in some Western European countries (Sustainable urban development [URBAN], 2000), notably Great Britain (Dabinett, 1999) Sweden, Italy, Greece, and Spain (Chorianopoulos, 2002), as well as some developing areas like India, Tanzania, Nigeria, Vietnam, and Pacific Asia (cf. Chakrabarti, 2001; Armstrong, 1993; Ayeni, 1991; Bolay and Du, 1999; Douglass, 2000).

1.3.2 City Management

To the extent that studies have focused on managerial and policy-making issues in urban crises, they link up to a strand of research within urban studies that focuses on city management and the politics of urban development. A number of studies have focused on the role of leaders and leadership in metropolitan areas, with regard to urban political life in general (Kearney and Scavo, 2001; Krebs and Pelissero, 2001; Norris, 2001). Others have focused on leadership in specific policy fields, such as education, health (Shafi, 1994), and poverty (Uddin and Jones, 2000; Singer, 1994).

1.3.3 Urban Cultural Geography and Sociology

Despite the fact that densely populated areas with a rapid growth rate have always posed higher risks and complex problems, it was not until after World War II that the notion of urban crisis really became substantiated and came to be associated with an explosive mix of cultural, racial, and economic issues. The spectacular growth of cities and the development of conurbations with multi-million populations undoubtedly proved to be one of the most striking phenomena of the second half of the twentieth century. These centers, it has become apparent, generate both negative side effects such as pollution, criminality, and delinquency, and positive effects such as job opportunities and cultural flourishing.

With the growing economic disparities and racial tensions in Detroit after World War II, the Western world came to know the growing racial tensions and economic problems in the inner cities as "America's urban crisis" (see for instance Sugrue, 1999; Nelson, Schwirian, and Schwirian, 1998). The Watts riots in 1965 came to epitomize a new type of violent urban crisis and the reprecussions it could have (Crump, 1966; Governor's Commission, 1965). In the early 1990s, America and the world had to reacquaint itself with the notion of "urban crisis" as the 1992 Rodney King incident in Los Angeles sparked violent inner-city riots that spread across not only the United States but also to metropolitan areas in other countries (Staples, 1994; *Business Week*, 1992; United States Congress, 1992; *The World & I*, 1991).

One of the underlying causes of future urban crises, some researchers argue, is the combination of demographical changes with a concentration of particular demographic strata, which is occurring in so called megacities (Dror et al., 1999). Another aspect of the demographic problem, is connected to the growing population of elderly in many Western societies. This increase in the oldest population stratum is a phonomenon that has been observed all over the world, but has become a particularly pressing problem in countries like Japan. The older urban population constitutes a particularly vulnerable part of the population that will be gravely affected in crises and this vulnerability increases with an increasingly stressed and frail urban infrastructure (Castenfors, 1996). Scholars like Castenfors have also pointed out that elderly people and their special needs in crises have typically been neglected in crisis studies but should be recognized as a special victim group because of their growing presence in urban areas.

1.3.4 Crisis Management in Urban Areas

The study presented in this book is in part an effort to examine and explore the concept of urban crisis and what some of the peculiarities and constraints associated with it may be. The crisis management side of the literature on urban crisis has been heavily lopsided in its focus on risk factors in the urban environment and various social justice aspects of the distribution of risk in urban areas (see for instance Morello-Frosch, Pastor, and Sadd, 2001). Very little has been written about crisis management in larger metropolitan areas and the particular challenges that these geographical and cultural areas may pose for crisis managers and policy makers, despite a few notable exceptions such as Jacobs (1997); Boulle, Vrolijks, and Palm (1997); and Rosenthal et al. (1994).

Most of what has been written pertaining to crisis management in urban environments to date (2002), has been in the form of case studies. These case studies seem to have been conducted without a particular focus on the urban or metropolitan setting as posing particular problems for policymakers, i.e., focusing on this setting as an independent variable affecting crisis management outcomes as the dependent variable. Again, some notable exceptions exist such as Miller (2001), writing about crisis paralysis; and DiPasquale and Glaeser (1998) on the economics of urban unrest in connection to the 1992 L.A. riots

A number of studies have focused on crisis management challenges associated with urban risks and accidents. Studies like these have dealt with, among other things, the management of hazardous materials transports and industrial contamination, and the impact and management of integrated high-tech communications systems in urban areas (Graham, 2002; Zimmerman, 2001; Warren, 2001; Coates, 2001; Widmayer, 2000).

In the international crisis literature it is suggested that the utilization and development of technological systems and risky technology is one of the prominent up-and-coming sources of crises in general, and urban crises in particular. Scholars making this argument highlight the increasing interdependence built into the systems we design and the increasing insecurity of these systems. The theory of "normal accidents" (Perrow, 1994) is one of the theoretical contributions that directly address this problem. One of Perrow's arguments is that regardless of the effort we make to try and predict accidents, they will continue to occur mainly because of the interactive complexity of tight couplings in the high-risk systems we have created. These accidents are normal in the sense that they are a direct result of the calculated risks that are built into system designs. The current design and utilization of high-risk systems make accidents inevitable (Perrow, 1994). The management problem that Perrow highlights is the decreased flexibility and maneuverability that is created with tightly coupled and interdependent urban systems. Smaller problems become harder to limit or to isolate early on in these systems—something which in turn threatens to result in cascade effects bringing down parts of the system or the system as a whole (Perrow, 1999; Wildavsky, 1988).

Perrow's theory of complex and integrated systems serves as an example of the kind of theory that may become highly relevant as the field of urban crisis management develops. It suggests that large urban areas and metropolises can be seen fruitfully, from a crisis management perspective, as webs of integrated and complex systems. From this theoretical perspective, a crisis in a metropolitan area can be understood as causing ripple effects that impact all levels of society and that can threaten to change the foundations of society, in a way that makes it difficult to return to any point of origin or normalcy within that urban community (t'Hart, Rosenthal, and Boin, 1999). Case study examples where urban crises have demonstrated these types of characteristics are the Heizel Stadium disaster (Rosenthal, Charles, and 't Hart, 1989), the assasination of Prime Minister Olof Palme (Hansen and Stern, 2000), the Gothenburg fire disaster (Hagstrom and Sundelius, 1999), and the Sakhalin earthquake disaster (Porfiriev, 2002).

Some of the specific problems associated with urban crises are intrinsically linked to the often unique physical character and form of the city itself. The structure of large cities, for instance, pose particular logistical strains such as heavy transportation and transportation of hazardous materials (Castenfors and Svedin, 2001). Transports of hazardous materials through densely populated city areas constitute a threat that has been manifested (cf. Castenfors and Stenstrom, 1998; Castenfors and Svedin, 2001) and internationally debated frequently over the last few years.

Others have focused on the operational management side of urban crisis managment. Problems in urban environments, some of these scholars argue, tend to accumulate and will eventually threaten to lead to a greater collapse. Policymakers and crisis managers in these situations have to manage a large number of acute and threatening situations occuring at the same time or in close connection to each other.

Other urban management problems that have been highlighted in relation to crises are the great amount of resources that an urban crisis may demand from rescue services. Hard questions about priorities may become pertinent when several incidents tax the same resources and this may require substantial coordination between districts and levels of authority ('t Hart, 1999).

Another operational management problem of urban crises is so called disaster tourism (Rosenthal and 't Hart, 1998). This becomes a practical problem, particularly when disasters happen in densely populated areas where large crowds of spectators and volunteers gather quickly. Several case studies of crises that have taken place in urban areas testify to how this pheonomenon can hamper and become a real obstacle in rescue efforts. The responses of the police and the rescue services in the cases of the Bijlemer air disaster in the Netherlands and the Gothenburg fire disaster in Sweden, were notably affected and in some cases complicated by the presence of large numbers of spectators and volunteers (Rosenthal et al., 1994; Hagstrom and Sundelius, 2001).

A recent and salient example of a large scale urban crisis occurred September 11, 2001, with the attack on the World Trade Center in New York. This event really exemplified the above-mentioned problems on a large scale and at a level of operational complexity that pales in comparison to many other types of disasters. The analysis of the New York response to September 11 illustrates this well with topics ranging from the psychological impact on victims and response units (Miller, 2002; Underwood and Kalafat, 2002; Water, 2002), to urban technological and infrastructural impacts (Davies, 2002; Haimes, 2002; Skinner, Schear and Katz, 2002), mediatization effects (cf. Pennebaker and Lay, 2002), and sociocultural changes in peoples perception of place and space (Halbert, 2002).

Urban crises, as exemplified by this study, seem to distinguish themselves from other types of crises in two interesting ways, both with regard to the sources that contribute to these crises, and the effects that this type of crisis has on society. The decay of older cities, the deterioration of the urban environment and the wearing down of old physical infractructure, can be picked out from the urban planning litterature as potential sources of crises. Urban crises seem furthermore seem to often be tied to complex and risky conditions inherent in complicated infrastructural systems and organizational practices. All in all, studies that embrace a more all-encompassing approach to the challenges of urban crisis management seems to be conspicously lacking.

1.4 CONTEXTUAL PROCESS ANALYSIS

It is generally very difficult to comprehend an organizational response to a critical incident without reference to the prior history of the entity in question. This prior history creates psychological and social dispositions toward perceiving and responding to problems in particular ways (cf. Steinbruner, 1974; Allison, 1971). Prior experience permeates the form of the current organisation, the degree and character of pre-crisis readiness and asset allocation, the set of available historical analogies, and the norms and values internalized by key individuals and/or encoded in group or organisational cultures and subcultures (cf. Lebow, 1981; Hollis and Smith, 1990; Katzenstein, 1996). Knowledge of broader societally-based cultural patterns is often indispensable (Sampson, 1987; Vertzberger, 1990; cf. Green and Shapiro, 1994:27).

Not only knowledge of the past is necessary, but also knowledge of the contemporary political and administrative scene. What are the most important issues which have been preoccupying players just prior to and during a crisis? What other issues compete for time and attention with the crisis issue or are cognitively available as a source of inspiration for problem defining and solving? Similarly, good knowledge of the institutional setting is essential.

This includes both the formal rules and the prevailing political-administrative practices, particularly the distribution of mandates and authority for crisis management.

Thirdly, our approach focuses attention on the pivotal role of subjective problem definitions and value articulations (George, 1980; Keeney, 1992; Soltan, 1998) in decision making. Problem definitions and value prioritisations cannot be assumed to be objective and incontestable. Rather they are the product of highly subjective and contingent cognitive and organizational processes. Problem definition, in crisis as in other types of decision settings, is an iterative process in which preliminary understandings of the problem and of possible solutions confront each other until a predominant view or constellation of views emerges (George, 1980; March and Olsen, 1989). Even then, problem definitions and their associated value constellations may change, sometimes dramatically, over the course of a particular crisis. Reconstructing the decision-making process in a manner which attends to changing patterns of perception and participation over time, and which can provide answers to the questions "what did they know?" and "when did they know it?" is an essential point of departure for crisis analysis.

For this reason, a bird's eye view of crisis decision making which treats a particular crisis as a single integral type of policy predicament is less useful. For one, such an approach tends to diverge sharply from the experiences of the participants on the ground. These officials are likely to be acutely concerned with how to handle specific and immediate situations perceived as relevant to their areas of responsibility: "one damn thing after another." The complex division of labour and limited degree of problem integration maintained by the highest levels in a policy-making system contribute to problem fragmentation. Thus, "[t]he decision process does not appear to be particularly oriented towards long term outcomes; far more attention is paid to the problem of what to do next." (Anderson, 1987:298). Particularly in the tense context of a major crisis, there is a tendency among policy-makers and organizations to focus on the next step rather than on the longer path ahead (George, 1986:542).

Viewed in this way, crisis decision makers are preoccupied with successive emergencies—fighting fires rather than thinking about preventing them. Instead of dealing with a single broadly-defined problem, they commonly experience a series of tactical sub-problems of varying degrees of urgency. It is clearly important for crisis analysts to develop methods suited to capturing this key aspect of crisis response. These should enable us to isolate those crucial "moments of choice" (also termed "decision occasions") that taken together have shaped the official (or communal) response to an emergency in a decisive manner. When does the crisis begin and end? For many crisis cases, this is a relatively simple matter. The analyst attempts to identify the development (the "impetus") which serves to create crisis levels of stress for the decision makers and institutional structures under study. For example, the simultaneous terror attacks on the World Trade Center and the Pentagon on September 11, 2001 plunged New York City and the U.S. federal government into crisis.

Where escalation or de-escalation is gradual, chronological delimitation may be more difficult. For example, many natural or technological disasters are characterized by a typical pattern of rapid escalation and gradual de-escalation. Furthermore, while some crises are managed in a fashion that leads to relatively rapid recovery and political-administrative closure, others drag on as ongoing political traumas that may plague a polity for years or decades after the critical events. Focused analysis of such cases falls outside of the framework of analysis developed in this study (cf. Hansén and Stern, 2001).

Having established, at least in preliminary fashion (obviously such delimitations are often revised subsequent to further empirical research and reassessment) temporal parameters delimiting the crisis, the analyst constructs an empirical narrative. This is accomplished using as much of the full spectrum of available sources as is practically feasible. Generally speaking, this will include most of the following published government documents, archival materials (if available), reports from commissions of inquiry, testimony from the congressional or parliamentary record, newspaper and journal articles, transcripts or audio-video archives documenting radio and television news broadcasts. When they are available, previous scholarly articles and monographs are likely to be very useful. These primary and secondary sources may be complemented by interviewing, to the extent that access to participants and stakeholders is forthcoming. Through a synthetic process of source comparison and criticism a narrative account of the crisis can be constructed. It should be noted that this step is particularly difficult, demanding, and time-consuming.

In breaking up the incident into its component decision occasions, the task is to attempt to isolate the most important "what to do next" sub-problems. The premise of the process dissection strategy employed in this study is that decision occasions are generated as reactions to some kind of stimulus, or "impetus." The impetus to decision may come from inside or from outside the government. It may result from the occurrence of a dramatic focusing event such as a major accident (e.g. the explosion of one of the nuclear reactors at Chernobyl), dramatic changes in economic indicators (e.g., a stock market crash), an intelligence warning, or publication of scandalous information by the media (cf. Steinbruner, 1974:146; Kingdon, 1984:206–207). The principle of disaggregation is to attempt to: (a) document an impetus, (b) reconstruct the subjective problem definitions of the key actors and stakeholders, (c) trace the formation of decision units which engage in response to this particular impetus, (d) identify the option or options discussed by this unit or units, (e) reconstruct the process resulting in (in)action, and (f) examine implementation.

It should be kept in mind that decision occasions may unfold in a fashion which diverges from normative models of rationalistic decisionmaking. For example, Kingdon (1984:215; cf. Anderson, 1987; Allison and Zelikow, 1999) suggests that: " participants do not first identify problems and then seek solutions for them; indeed, advocacy of solutions often precedes the highlighting of problems to which they become attached." Kingdon introduces the concept of policy windows, which are opportunities for linking policy problems with policy alternatives, created by the rhythms of political life. In some sense then, a decision occasion may be viewed as an open policy window.

The particular composition of the decision unit which ultimately engages to cope with the issue in question is the result of the interplay of many factors. Real-world constellations of decision responsibility in crises tend to diverge dramatically from those described in abstract organisational charts or mandated in constitutional arrangements. In most states, constitutional stipulations or crisis response regulations provide guidelines about who should be involved in what. Such institutional features do, of course, affect the shaping of the decision unit in a particular case. However, formal rules alone do not determine the composition of a decision unit. Decision units are instead formed in a complex interplay between the codified requirements of governing, other embedded features such as (often informal) institutional rules and practices and contextual factors; they are formed on the basis of many considerations.

Contextual factors such as the amount of time available, the seriousness of the problem, and the physical availability of officials play a role. The substantive nature of the impetus determines to a large extent which organizations may make a claim to participation and forms a basis for selecting technocratic expertise (cf. Rosenthal and 't Hart, 1991). Political sensitivity and their level of preoccupation with other pressing issues, help to determine the level of interest and participation among political leaders. Personal relationships among decision makers often play a role.

This approach to crisis response analysis devotes considerable effort to tracing the interplay between situations, actors and institutions prior to the point of choice. This is inspired by the belief that an institutionally embedded structuration process shapes many aspects of the ultimate decision long before choices are made. This view of crises does not deny the role of individuals, but it takes note of the structural forces shaping their thoughts and actions. Each decision occasion is delimited by the subjectively experienced arrival of a major "what to do next" question. Through a sequence of what may appear to be minor steps on the part of the engaged decision units, the policies are prepared to cope with particular segments of the broader issues at hand.

The final element of the inquiry into each decision occasion is consideration of the implementation phase. The transition from policy making to policy output is often problematic; unanticipated logistical difficulties, environmental changes, communication and coordination problems, ambiguous directives, and deliberate resistance are among the many and varied factors which can intervene between the authoritative commitments made by high-level decision units and the resulting actions by subordinates or field personnel (Allison, 1971; Pressman and Wildavsky, 1984; Smith and Clarke, 1985; and Sabatier, 1993)

By breaking a historical crisis experience down into a series of decision occasions, complex decision-making processes can be traced in relatively close approximation to the reality perceived (in fragmented fashion) by the participants themselves. Of course one must be selective in identifying the occasions studied in depth. Herein lies a virtually unavoidable element of subjectivity. It is no simple task to try to reconstruct the operating conditions facing the decision makers at the time. However, through intensive empirical analysis of the available source materials, plausible interpretations may be developed. While such interpretations are inevitably imperfect and to some extent contestable, they promise to advance our thinking about crisis decision-making processes.

1.5 THEMATIC ANALYSIS

As we have argued, process dissection is an extremely useful (and not infrequently indispensible) tool for the analyst who wishes to develop insight into the dynamics of crisis decisionmaking in real world cases. However, it is by no means the only arrow in the analytical quiver. In the Crisis Management (CM) Europe research program of which this book is a part, comparative thematic analysis is used as a complement to process dissection. Two types of comparisons are particularly useful. One possibility is to compare thematically across the occasions for decisions identified in a particular case. Another is for the analyst to compare the relevant theme findings for a given case, with those identified in previous cases which have been studied using the CM Europe approach.[2] The following are the research themes which have been highlighted by the CM Europe program:

Preparedness, Prevention, and Mitigation: This theme focuses on the extent to which crisis managers and their organizations are prepared to respond to extraordinary events. Have they experienced serious crises before? Have they cultivated an "it can happen here" mentality and prepared themselves psychologically for the rigorous demands for crisis management? Do plans and infrastructure exist and are they adapted to the exigencies of the situation? Are decision makers able to identify potential threats and act to prevent them from escalating? Are windows of opportunity for limiting damage or making gains exploited?[3]

Leadership: This theme focuses on the leadership styles and types exhibited by key actors in the crisis. Leadership may be operative—in the sense of taking charge of the crisis management activities and coordination. It may also be symbolic—in the sense of showing through word and deed that the leading figures are participating in the crisis and empathizing with those who are at risk or who have already suffered losses. Leadership may be concrete and personal, or more abstract and distant. In deliberations, leaders may be directive/hierarchical or facilitative/collegial. Leadership should be seen as relational, inextricably linked to followership and mediated by factors such as power, affect, culture, organizational structure, access to expertise and context. Finally, leaders vary in their propensities to micromanage or delegate when it comes to critical decisions (see e.g. Bass, 1997; Gardner, 1995; Hermann and Hagan, 1998).

Decision Units: This theme focuses on the question of how and where decisions are made in the complex institutional systems which typically engage in crisis management. For example, crucial decisions can be made by single individuals, tandems or dyads, small groups, organizational networks, etc. Similarly, decision units may be located and play various roles within a given political/administrative (or corporate) system. Decision units may be strategic or operative in nature and may be located at local, regional, national, or supranational (e.g. EU or UN) levels. Decision units may vary considerably in terms of their composition, mode of operation, and placement across the course of a given crisis—it is not uncommon for shifts upward (upscaling) and downward (downscaling) to occur as part of complex politico-administrative escalation and de-escalation processes (see e.g. Snyder, Bruck, and Sapin, 1963; Hermann, Hermann, and Hagan, 1987; Stewart, Hermann, and Hermann, 1989; Rosenthal, 't Hart, and Kouzmin, 1991; Stern, 1999). Finally, what are the criteria for determining which individuals and organizations will be included in (or excluded from) the decisional "loop" and the decision unit?

Problem Perception and Framing: This theme focuses on the subjective and socially constructed nature of crisis decision making. Actors act not on uncontestable and objective knowledge of the situation but rather upon their perceptions and interpretations of what is happening (Snyder, Bruck, and Sapin, 1963; Sylvan and Voss, 1998; Stern, 1999). While problem framing often takes place on a semiconscious, intuitive level—especially by "naive" decision makers—problem-framing processes exert a profound influence upon choice. In other words, once a problem has been framed, many possible lines of action have already been discarded and strong propensities for and constraints upon action created. Framing is heavily influenced by cognitive and social structures and processes such as (historical) analogical and metaphoric reasoning, culture, context, organization, and information flow (Vertzberger, 1990; Larson, 1985; Khong, 1992). Why do particular actors perceive and frame problems as they do at various junctures of a crisis? Why do these problem representations change (or remain stable) over the course of a crisis and its aftermath?

Value Conflict: This theme focuses on the potential for tension or conflict among the various values at stake in a crisis situation. An integral part of problem framing is identifying the values, which are implicated in a given situation. This is often a demanding process and it is common for decision makers to overlook values embedded in a complex issue, if they do not engage in rigorous critical deliberations (see e.g. Steinbruner, 1974:16–17; Keeney, 1992). Crises, by definition have profound implications for fundamental values such as preservation of human life, sovereignty/autonomy, economic well being, democracy, rule of law, etc. More parochial values—such as the personal and political fortunes of individual officials and parties—are also involved. Good crisis performances make careers—poor ones break them. It is common for multiple values to be at stake in crises which means that decision makers often face painful and stress-inducing dilemmas as well as tragic choices (Janis and Mann, 1977). Decision makers may ignore or deny value conflict, which tends to generate unbalanced policymaking. They may recognize and accept value conflict—following Lenin's dictum that "You can't make an omelet without breaking eggs." They may choose to procrastinate and hope for the situation to improve. Alternatively, they may actively seek to resolve the conflict and find a transcendent solution, which adequately protects the key values at stake.[4] How, then, do actors cope with the value conflicts which tend to emerge in crisis situations?

Politico-Bureaucratic Cooperation and Conflict: This theme focuses on the issue of patterns of convergence and divergence, parochialism, and solidarity, among the actors and stakeholders engaged in a crisis. There are a number of documented dynamics, which tend to create pressures for cooperation and solidarity in crisis (e.g. the "rally around the flag" effect, leader attentiveness, and "groupthink").[5] However, there are also a number of countervailing tendencies. Crises often present particularistic risks which may induce political or bureaucratic actors to engage in defensive behaviors, which may in turn antagonize other actors and lead to conflict. For example, following failures or setbacks, it is common for actors to play a "blame game." Equally importantly, crises present opportunities as well as risks and so actors may compete in seeking credit for their contribution (and in denigrating that of others). Finally, situational and contextual factors tend to be moderated by the nature of personal relationships within policy communities and the strength of national cultural norms opposing opportunism in extraordinary situations (see e.g. Rosenthal, 't Hart, and Kouzmin, 1991; Stern and Verbeek, 1998; Allison and Zelikow, 1999).

Crisis Communication and Credibility: This theme focuses on the relationship between crisis managers, the media, and elite/mass publics (see e.g. Nohrstedt and Tassew, 1993; Nordlund, 1994; Pearce, 1995; Regester and Larkin, 1998; Henry, 2000). In democratic polities, maintaining credibility and legitimacy with the media and the public is an essential task of successful governance in both crisis and more normal situations alike. Crisis managers enter crises with varying degrees of credibility and may gain or lose such over the course of the crisis. Actors vary considerably in their approach to crisis communication. Some take a defensive/closed stance, which can easily antagonize the media and cost credibility. Others take a more proactive/open stance and seek to maintain the initiative in providing information and establishing friendly relations with the mass media. Actors vary considerably in ways to the degree to which they coordinate crisis communication and in the information strategy and tactics developed. Similarly, some actors closely monitor how their messages are being received and act to correct problems, while others—focusing on other aspects of crisis management and distracted by stress—are oblivious to growing credibility problems. There are a number of recurring credibility "traps"—which involve creating gaps between words and deeds, between expectations and performance—that can cost crisis management dearly, as can the neglect of the symbolic aspects of crisis management (Stern, 1999:201–202; see also Hansén and Stern, 2001).

Transnationalization and Internationalization: This theme focuses on the tendency of crises to spill over national boundaries in an increasingly politically, economically, socially, and not least ecologically interdependent Europe. While some crisis-producing threats may arise within a single country, many actual and potential threats do not respect borders. Infectious diseases, natural and technical disasters, financial turbulence, and terrorism are just a few examples of such threats. Similarly, coping with contemporary crises tends to require trans- and international collaboration in order to deal with these cross-border threats (Buzan, 1991; 't Hart, Stern, and Sundelius, 1997; Buzan, Waever, and De Wilde, 1998; Steinbruner, 2000). Such cooperation may be ad hoc or institutionalized, bilateral or multilateral.

Temporal Effects: This theme focuses upon sequencing and synchronicity—temporal effects which may have a profound effect on crisis management. Sequencing refers to the path-dependent nature of crisis decision-making. Choices made early in crises tend to constrain the possibilities for later action and steer the crisis management along a particular trajectory which may be difficult to change (see e.g. Levy, 1991; Sundelius, Stern, and Bynander, 1997; Billings and Hermann, 1998). Often feedback, especially negative feedback, from early decisions will require further decisions—which together form a sequence. Synchronicity refers to the tendency of simultaneous events to influence each other via psychological and organizational mechanisms such as "availability," opportunity cost, cumulative stress, and distraction (Snyder, Bruck, and Sapin, 1963; Haney, 1997; Stern, 1999). Synchronicity may appear within a single crisis (when multiple problems must be solved at the same time), between two simultaneous crises (as in the cases of Hungary/Suez in 1956, and Watergate/ War in the Middle East in 1973), or between a crisis and other highly prioritized coincidental activities, such as elections, state visits, or crucial legislative negotiations.

Learning: This theme focuses upon the extent to which actors are capable of analyzing their experiences and using the results as basis for change. As noted above in the discussion of problem framing, actors may attempt to use "lessons" from past experiences (encoded as historical analogies or as experientially-based "rules of thumb" as a guide for current action. Similarly, actors may respond to positive or negative feedback regarding performance during a crisis, by drawing lessons and modifying beliefs and practices. Actors commonly attempt to reflect upon crisis experiences after the fact, draw lessons for the future, and formulate reform projects on the basis of interpretations of crisis experiences. Crises present considerable opportunities for learning, but post-crisis learning attempts are often distorted or derailed by a variety of typical social and psychological dynamics (see e.g. Lebow, 1981; Levy, 1994; Breslauer and Tetlock, 1991; Lagadec, 1997; Stern, 1997).

1.6 OUTLINE OF THE STUDY

The study is structured in the following fashion:

Chapter 2 sets the stage; a depiction of the institutional, legal, and political/cultural contexts is presented. Background on relevant agencies and departments in New Zealand will be provided.

Chapter 3 focuses on the acute phase of the crisis and the return to normalcy. Key problems of decision making, coordination, and communication facing the crisis actors will be identified and the exercise of authority analyzed. Which were the most important actors and stakeholders at various stages of the crisis? Was the locus of decision relatively constant or did it shift a number of times during the incident? Were the response and recovery phases characterized by solidarity, public confidence, and cooperation, or by blaming, loss of legitimacy for engaged actors, and conflict? The answers to these and similar questions are addressed in the third part of the study.

Chapter 4 consists of a series of discussions focusing on particular themes relevant to the Auckland case. Each one isolates a particular theoretical and or practical issue identified in the crisis literature and applies it to the Auckland case. The following themes are addressed in this study:

- Crisis Development
- Problem Framing
- Leadership
- Organizational Cooperation and Conflict
- Crisis Communication and Mass Media
- Intelligence Gathering and Information Processing
- Practical Issues and Stress Management
- Symbolic Politics versus Technical and Managerial Logic
- Experts and Decision Makers
- Crisis Management in an Urban Setting
- Sequencing and Synchronicity
- Learning

When appropriate, the Auckland findings are complemented with comparative case illustrations from the CM Baltic/Europe case bank or from the international literature. The study concludes with a summary of the most important practical lessons derived from the analysis of this case, lessons likely to be of use in preparing for or responding to future crises.

NOTES

1. During a two-week trip to New Zealand in late 1998 more than twenty interviews with participants and knowledgeable observers were conducted (see the bibliography of this study for a more complete list).

2. The following section builds upon Sundelius, Stern, and Bynander (1997) and Nohrstedt and Sundelius (1999) and is taken directly from Stern and Sundelius (2002).

3. On crisis prevention and mitigation, see, e.g., Ender and Kim (1988), Waugh (1988), and Lund (1996). For more skeptical examinations of the potential for risk elimination and accident prevention, see Wildavsky (1988) and Perrow (1999).

4. On value conflict, see George (1980) and Farnham (1998).

5. See, e.g., Rosati (1981) and Janis (1982).

2

Public Sector Reform, Electricity Policy, and Crisis Preparedness in New Zealand

2.1 A COUNTRY IN TRANSITION

New Zealand, or Aotearoa in the indigenous Maori language, was officially created in 1840 with the signing of the Treaty of Waitangi by Lieutenant-Governor Hobson representing the British crown and by chiefs representing the Maori nation (Sullivan, 1997; Bishop, 1998). While the treaty was purported to be a charter for power sharing between the parties and allow for Maori self-determination on their indigenous lands, the documented experience and voice of a great many attest to the fact that the arrangement has resulted in domination and discrimination of the Maori by the Pakeha, or settler population, resulting in unequal opportunity and social dislocation (ibid.) The colonial history of New Zealand[1] is important to understanding current political developments in the country, as the issue of Maori sovereignty has become topical in recent years (Durie, 1997).

From 1852, with the signing of the New Zealand Constitution Act, further reinforced by the Letters Patent of 1917, until well into the twentieth century, the Governor-General of New Zealand, representing the Queen of England, wielded great power in the colony, although it was seldom exercized contrary to local colonial governing interest (Ladley, 1997). This tradition continued, and in 1983, the signing of a new Letters Patent and in 1986 the passing of the Constitution Act, effectively curtailed the powers of the Governor-General to block or deny the passing of legislation, rendering the role of the monarchy in New Zealand as "largely ceremonial and symbolic" (ibid.:51).

In strictly legal terms however, the Governor-General continues to wield great reserve powers, as that official is able to act contrary to the advice of a Prime Minister who has lost the confidence of the legislature, and to deny activity of a government should it be perceived to be acting in serious violation of constitutional convention (ibid.:57–59). Further, the Governor-General's signed approval is still required for passing of legislation, although a denial to do so would cause, as would the invoking of other reserve powers, a constitutional crisis. Accurately stated, the "formal governing executive body of New Zealand is not the Prime Minister and cabinet, but rather the executive Council, which comprises Her Majesty's Governor-General in meeting with the ministers of the Crown" (ibid.:52), although "it is well understood that the reality of governmental power lies in the meetings and decisions of the cabinet" (ibid.:52) which the Governor-General "rubber stamps."

2.2 PUBLIC SECTOR REFORM

Extensive reform of the economy and public sector have taken place in New Zealand since 1984, on a scale described by some as "revolutionary" (Spicer, Emanuel, and Powell, 1996:xv).[2] "With only some exaggeration, observers compare the thoroughgoing programme of market liberalization, deregulation, corporatization and privatization to the dismantling of communism in Eastern Europe. The scope of the restructuring defies easy summary" (Nagle, 1998:229).[3] These radical changes have been attributed to responses to serious macroeconomic problems in the country with roots stretching back several decades. Major contributing factors include recessions caused by the dramatic reduction in the price of wool in 1967 and 1968, an export on which the small trading nation was highly dependent, the end of their favoured trading relationship with Britain and the world oil crisis both in 1973 and a number of misdirected domestic approaches, notably the ill-fated "Think Big" energy scheme of the National government under Muldoon, embarked on in the late 1970s and early 1980s (see Dalziel and Lattimore, 1996). The first reforms were implemented during the Labour Government's two consecutive terms of office from 1984–1990[4] and were continued under the National Government, which dominated the legislature in the 1990s. Following an Economic Summit Conference called by the Labour Government two months after its election in September of 1984, "and subsequent conferences devoted to Maori economic development (October 1984) and to employment (March 1985), the government had a clear mandate to initiate a process of economic reforms" (Dalziel and Lattimore, 1996:25). New Zealand, a nation which had been a committed welfare state since the mid-1930s thus began a process of great socioeconomic change (McClintock, 1998). The government "introduced major structural reforms in the ownership, organisation and management of government enterprises, the core public sector, local government, ports, airports, retail electricity distributors, health services, science and education" (Spicer, Emanuel, and Powell, 1996:3). Some of the most notable changes enabled under the program of economic reform included

> the opening up of the economy to international competition and investment; the removal of monopolistic advantages, regulations and subsidies from virtually every sector; a shifting of taxes from income to consumption and a flattening of income tax rates; reduction of many state benefits, often by targeting (means-testing) or user charges; a decrease in state employment of about 80,000, including 29 per cent in the core public service; the hiring and firing of government executives based on performance contracts . . .; and a labour relations law [the Employment Contracts Act 1991] based entirely on the principle of individual contracts (Nagel, 1998:229–230).

A large public sector in New Zealand was radically transformed, first through commercialization, whereby economic efficiency was prioritised in the commercial activities of government, followed by corporatisation of government trading activities, whereby departments or enterprises of government involved in trading were transferred into State-Owned Enterprises (SOEs) (Spicer, Emanuel, and Powell, 1996). Prior to these changes, government trading enterprises accounted for about "20 percent of gross investment and over 12 percent of GDP" in New Zealand, a figure comparatively high in OECD terms (Boston, 1998:33). Following the change, SOEs in New Zealand operate in a fashion similar to that of private sector corporations. The extensive corporatization in New Zealand has been highly controversial. The lack of government intervention allowed under the new legislation has been criticized as well as heralded for removing the possibility for political and social factors to intervene in consideration of the economic (Dalziel and Lattimore, 1996; Spicer, Emanuel, and Powell, 1996; Nagel, 1998). However contested, they were undoubtedly big in scope and swift in pace. The number of public servants in New Zealand was reduced from around 88,000 when the corporatization program was initiated, to around 33,000 by the late 1990s (Boston, 1998).

In line with their corporate profile, SOEs are required to produce an annual Statement of Corporate Intent (SCI),[5] "an important document which constitutes a performance agreement between the shareholding Ministers and an SOE board. Its value [according to its proponents] lies in its clear commercial focus on such issues as definition of core business, performance targets, dividends and accounting policies" (Spicer, Emanuel, and Powell, 1996:184). SCIs must comply with the SOE Act which explicitly states that SOEs must operate as successful businesses, ensuring both profitability and efficiency comparable to businesses in the private sector (ibid.) The State-Owned-Enterprise (SOE) model was then applied "more widely as an archetype for other restructuring in New Zealand in science and technology, health, local government, ports, airports and retail electricity supply" (Spicer, Emanuel, and Powell, 1996:3). The elaboration of this model to certain traditionally non-profit sectors of the economy, such as the health sector, has resulted in questions of accountability and punctuated class differences.[6]

The attempt to corporatize the Ministry of Works was particularly fraught with difficulties, leading to an ultimate splintering and reduction in services.[7] The Ministry of Works and Development was closed on April 1, 1987 and under a piece of legislation called the Public Works Act, an SOE called "WORKS" was formed out of the former Ministry's commercial division (Spicer, Emanuel, and Powell, 1996). Under its new corporate structure, the CEO restructured the organization into "a corporate office and four main operating divisions organised around four broad lines of business: consultancy services, civil construction, property services and computing services. Each of the divisions was to be treated as a full profit centre" (ibid.:102). Due to problems with the corporate transition, attributed to conflicting "organizational cultures" and their lack of profitability, WORKS downsized further and divested through "management buyouts" both the computing services and property divisions and established the remaining two divisions as "subsidiary companies" to enhance their marketability and comply with the legislative changes regarding SOEs (ibid.:103;128). The Civil Defence Act of New Zealand (1983) was amended in 1988 (Section 19.2.c was repealed) in order to reflect the absence of the Commissioner of the Department of WORKS from the National Civil Defence Committee (Government of New Zealand, 1983; 1988).

2.3 ELECTRICITY REFORM

The electricity sector, like many of the sectors in New Zealand during the past fifteen years of broad economic reform in the country, has undergone—and continues to undergo—extensive and often controversial reform. The generation and transmission of electricity were under the jurisdiction of a central government department, the Ministry of Energy, until April, 1986 (Ministry of Commerce, New Zealand, 1999b). With efficiency in mind, the government created at this time the State-Owned Enterprise (SOE), the Electricity Corporation of New Zealand (ECNZ), to own and control the electricity division (the New Zealand Electricity Division (NZED)) of the Ministry of Energy (ibid.)[8]

Following establishment of the ECNZ, "a fundamental realignment of the organisation taken over from the NZED" took place (Spicer, Emanuel, and Powell, 1996:96). Operating divisions were established in four separate areas—Production (electricity generation), Marketing (marketing and sale of electricity), Transpower (high-voltage transmission network) and Power Design Build (design and construction) (ibid.) Once the subsidiary Transpower, was set up in 1988 to operate the transmission network, ECNZ confined its activities to the generation of electricity alone (Ministry of Commerce, New Zealand, 1999b). Design Power Limited, "established in 1987 as a distinct company with its own Board of Directors," operated as an independent subsidiary of ECNZ engaged in engineering consultancy (Design Power, 1999).[9] In a further corporatization effort, a Transpower Establishment Board was set up in 1990 to achieve separation of Transpower from ECNZ.

Following the Establishment Board's report in 1991, the Energy Sector Reform Bill, also addressing reform in the gas sector, was introduced to Parliament (Ministry of Commerce, New Zealand, 1999b). The Energy Sector Reform Bill entailed a number of wide-ranging reforms and ultimately resulted in the passing of several pieces of legislation. Distribution and supply of electricity in New Zealand was localized under a number of Electricity Supply Authorities (ESAs). Until 1990, these ESAs were under the jurisdiction of local government, following which they too were corporatized and privatized under local trusts (ibid.)[10] The Energy Companies Act of 1992 allowed for corporatization of the ESAs under share ownership plans, while the Electricity Act 1992 resulted primarily in deregulation of distribution and the creation of the Energy Efficiency and Conservation Authority (EECA) in 1992 and franchise restriction removals in 1993 and 1994 (ibid.)

In 1992, ECNZ "announced that the divisions of Marketing and Production would disappear and operate under the common name of ECNZ" as, according to the CEO, the culture of the old NZED had been sufficiently changed (Spicer, Emanuel, and Powell, 1996:99). Further justification forwarded regarding this change is that "with the separation of Transpower from ECNZ and its establishment as an independent SOE, it made little sense to have two or more internal units competing against each other for what was now, from ECNZ's perspective, external work involving external cash flows" (ibid.)

Advocates of corporatizing New Zealand's public sector emphasise the benefits of depoliticization attained through corporatization (Spicer,

Emanuel, and Powell, 1996). The electricity sector in particular has been held up as an example where this was necessary in order to counter "multiple, overlapping and sometimes conflicting functions and objectives [which] resulted in confused plans and hindered performance measurement and accountability" (ibid.:182). Replacement of the NZED with the SOE ECNZ in 1987, followed by the elimination of the Ministry of Energy in which the NZED had been part in 1989 were from this view, positive steps towards eliminating these multiple political and socioeconomic objectives and increasing economic efficiency (ibid.) The authors contend:

> The NZED provides an excellent illustration of the commingling of political, social and economic objectives. Although NZED's management had a record of good technical performance, the overriding orientation of this organisation was to assure a reliable supply of electricity to New Zealand households and industries. Over time this had led to the oversupply of electrical generating capacity, non-commercial pricing, a high degree of cross-subsidisation in favour of residential customers and a lack of focus on cost containment and efficiency (ibid.:182).

Concerns have been raised however, that the new, "lean" style of management adopted by the former NZED, the ECNZ, has resulted in an elimination of these former concerns to such a degree that security and supply as well as economic problems may arise. These concerns have increased following a number of deregulating changes, including the complete separation of Transpower from ECNZ in 1994, the split of ECNZ into two (forming an additional SOE called Contact Energy in 1995), and the establishment of a "fully competitive private sector wholesale market in 1996" (Ministry of Commerce, New Zealand, 1999b). Under these arrangements, Transpower is "the owner and operator of New Zealand's national electricity grid" (Transpower, 1999). The new SOE Contact Energy is responsible for generation of over 20 percent of New Zealand's total electricity supply (Ministry of Commerce, New Zealand, 1999a). A new mode of operation was introduced for the transmission of electricity by Transpower in 1997, whereby efficiency and cost-effectiveness were stressed in their new SCI and the initial steps towards making "customers" responsible for determination of the security of power supply were taken (Transpower, 1997). Up until this point, "Grid Security Policy" had been "set by Transpower with limited industry involvement" (Transpower, 1997). Transpower defines Grid Security Policy (GSP) as "a set of standards that applies to the real time supply of electricity and power system coordination. It covers such issues as the reserves of electricity required to meet unplanned power losses as well as the ability to interrupt supplies to meet such contingencies. It also covers the quality of electricity supplies in terms of the frequency and voltage of power system flows" (Transpower, 1997). The reliably regulated transmission of electricity in a power system is essential for preventing power disturbances. While generation, transmission and delivery of electricity are of course connected with change in one area affecting others, the problems which caused the Auckland power outage originated not in Transpower's national grid, but in the local Auckland grid owned and operated by the SOE Mercury Energy Limited. Most power outages of great magnitude are in fact, typically caused by generation problems, ie. demand problems and not problems with the distribution grid such as the case was in Auckland in 1998. International interest in the case has therefore been particularly keen.

A further measure aimed at opening up the electricity industry to increased competition, which has resulted in heightened concern about prices, and system security was the Electricity Industry Reform Act (EIR Act) "enacted by Parliament on 3 July 1998" (Commerce Commission, 1999). The act required, with few exceptions, all lines and transmission business to be separated from the generation and retail of electricity. Importantly, the act also allowed for the splitting-up of ECNZ (ibid.) The competition-geared changes allowed for 40 percent of the cornerstone shares of the SOE Contact Energy to be sold to Edison Mission Energy, an American multinational (Ministry of Commerce, 1999b). The remaining 60 percent of the less-influential Contact shares were "sold via a public share float" (ibid.) The Auckland electricity outage was used by both proponents and those opposed to the reform as justification of their position on the further breakdown of the ECNZ, which was more or less complete by the spring of 1999. On April 1, 1999 the ECNZ was split up into three competing SOEs: Mighty River Power, Genesis Energy, and Meridian Energy. "The three companies are capable of producing more than half of the nation's power needs" (Reynolds, 12/06/99).

Following the changes and the subsequent split, competition in the electricity sector in New Zealand has been fierce and a number of sales and takeovers have taken place. The competition, as referred to earlier, has been criticized on both financial and security grounds. Individual consumers and particularly smaller power companies are concerned about the unequal advantages of larger power consumers. Activists have argued against the split in the electricity sector on the grounds that those that would benefit from the change would be big electricity users rather than individuals and small consumers (Venter, 1998). The ESA on the other hand "welcomed the removal of ECNZ's market dominance, but not the forced separation of line and energy companies" (ibid.:2). While one of the arguments forwarded by proponents of deregulation and privatization of the electricity sector is reduction of costs for consumers, the experience in the United States indicates that prices for electricity have decreased only for large industrial consumers, with prices for individual consumers actually increasing. Moreover, the 2001 power crisis in California has brought the issue of reliability and security of privatised energy to new life.

According to the American government's Energy Information Administration (EIA), "[i]ndustrial customers are paying the lowest prices for electricity in 17 years" (Power Engineering, 1999:14). An administrator at EIA attributes the greater competition in electricity generation and the size advantage of industries to their negotiation of better prices with utilities. Despite the promises of cheaper electricity for all with deregulation, "[i]n contrast, the price of electricity to residential customers went up slightly between 1996 and 1997, reaching its highest level ever" (ibid.) Similarly, in the United Kingdom, which has been deregulating its electricity industry since 1990, "[t]he unit cost of electricity has decreased, but the price to the consumer has not" (*Consumer Reports*, 1998:6). Concern has been expressed within New Zealand that similar trends are already apparent.

What privatization-geared electricity reform definitely has and continues to result in in New Zealand is a dramatic increase in foreign ownership. Utilicorp, a multinational corporation based in Kansas "owns 37.5% of Auckland's Power New Zealand and 39% of Hamilton's WEL Energy through a joint venture with Todd Corp" (Ruth, 1998:21). Contact Energy, the first electricity generator to be split off from ECNZ in 1996 and corporatized has since expanded into retail sales of gas and electricity as well (Gaynor, 1999). As mentioned, the company, which has "approximately 25 per cent of New Zealand's generating capacity" and a "22 per cent national market" in retail electricity, has recently sold 40 percent of its shares to the American company, Edison International, giving the international cornerstone sway (ibid.) The Canadian-based company TransAlta owns an even larger percentage of the retail electricity market in New Zealand (ibid.) These changes indicate that certainly the reforms have generated a great deal of competition as they were intended to do, but the bulk of that competition takes place between multinational corporations that are not New Zealand-based.

Increasingly, questions have been raised regarding the security of New Zealand's electricity supply as a whole following these changes in the sector. The concern has been raised that in such a competitive market, generators of electricity may attempt to manipulate electricity prices by generating different amounts at different plants according to which location will yield a greater profit (Reynolds, 01/03/99). "The ability for companies to exploit the market increases as demand for power comes close to matching available supply" (ibid.) If there is valid cause for concern that companies may actually strive to achieve a tight balance between the demand for electricity and available supply in order to maximize profit, the results could be worrying indeed. Risks of blackouts are real in New Zealand as much of the electricity is hydro-produced and precipitation and other climatological factors influence both supply and demand. Moreover, demand for electricity is increasing in New Zealand at more than 2.5 percent annually according to a report published by the Centre for Advanced Engineering at the University of Canterbury (Sinclair, Knight, Mertz (SKM)/Leyland Consultants, 2000).

As under the new guidelines, no electricity generator is to control more than a third of the market, "industry consultants say that this means no electricity company will be perceived as having any responsibility to provide sufficient reserve generating capacity to ensure security of supply" (Reynolds, 01/03/99). The engineering firm SKM—formerly Leyland Consultants—has expressed doubt that the market will ensure security of supply and that the need to revert to governmental intervention in the field may be real (ibid.) This point was expressed clearly following electricity disturbances which occurred in 1992. A severe electricity shortage occurred during that year when the catchment areas Huntly hydroelectric station relies upon were much lower than they usually were due to arid weather. A Commission of Inquiry following this crisis observed that the problem with the lake levels was not responded to via electricity price controls, which exacerbated the problem.

The Auckland power outage in contrast occurred in Mercury Energy's grid near the point of delivery and did not originate at the point of generation or in the national distribution grid. The system is however interconnected and concerns regarding the overall administration and operation of the industry are common. The New Zealand Electricity Market (NZEM) is administered by a subsidiary of the foreign multinational, the Rand Merchant Bank of Australia (RMBA). "In New Zealand RMBA's wholly owned subsidiary the Marketplace Company, M-co. is the operator of the wholesale electricity market. M-co. is the administrator, clearing and pricing manager of NZEM and a central provider to the industry" (RMBA, 1999). Companies involved in the generation and distribution of electricity in New Zealand participate voluntarily in this market and are expected to conform to rules under a Metering and Reconciliation Information Agreement (MARIA) administered by the Maria Governance Board (MGB) within the NZEM (M-co., 1999).

2.4 LOCAL GOVERNMENT REFORM

Local government in New Zealand falls under one of four categories: regional; territorial, community, and "ad hoc" (Bush, 1997). The territorial bodies are the fifteen city councils and fifty-nine district councils which are the frontline multipurpose deliverers of services" (ibid.:118). These councils "relate to the twelve regional councils, which, in conjunction with the four Territorial Local Authorities (TLA), combining both regional and local status, have a largely regulatory role" (ibid.:118). There are an additional fifty other TLAs in New Zealand communities, supporting along with the others, currently 154 community boards (LGNZ, 1999), and acting "primarily as links between the community and its local body" (Bush, 1997:118).

Local government fulfills many key functions in New Zealand. The importance of local government in New Zealand has been attributed to the fact that it is not a federal state with intermediary provincial or state governments, and thus, local government must assume responsibilities not administered at the central level (ibid.) An authority in the area of local government critically observes that "local government has been the obvious 'rubbish bin' in which central government can dump unwanted responsibilities. Thus despite the heavyweight areas of education, social welfare, and (conditionally) health lying beyond its realm, there are approximately sixty separately identifiable functions which are intra vires to local government" (ibid.:118).

Central government circumscribes local government operation under the Local Government Act of 1974, and the Local Government Amendment Act (LGAA) of 1989 (Bush, 1997:118; Hayward, 1997:414). As part of the broader reform package, the 1989 amendment resulted in changes in local government similar to those at the national level. Local government, like central, is now run like a business according to managerialist principles, with Chief Executive Officers (CEOs) and government officials maximizing utility wherever possible. Further changes are planned for upcoming years. Local government in New Zealand is legislatively committed to functions "such as the preparation and maintenance of planning, and civil defence schemes," which includes managing the natural and physical environment and the important function of planning under the Resource Management Act, 1991 (Bush, 1997:118). While not legally required to fulfill numerous other functions, expectations of central government and communities are that local government provide comprehensive service to their communities (ibid.) Significant changes with respect to civil defense have taken place in accordance with the amendment which are discussed in greater detail in chapter 3. Further, a number of other less all-encompassing amendments to the 1974 act have been implemented which have significant impact on certain regions such as that of Auckland which are discussed in the following section.

There are several administrative bodies, which serve to link the central government to the local bodies of government in New Zealand. There is a Minister of Local Government on the central government level as well as the organization, Local Government New Zealand (LGNZ) which is an elected body comprised of officials from different branches of local government in New Zealand (LGNZ, 1999). This body represents the interests of local government to the national government (ibid.) LGNZ is "governed by a 'National Council' of 14 persons elected to National Council by geographic groupings of councils (zones) or by groupings of types of councils (sector groups)" (LGNZ, 1999). Membership to LGNZ is voluntary and supported primarily by subscription, but as central objectives of the National Council of LGNZ are provision of a "collective voice of [local] councils at a national level," and "promotion of collective interests, information sharing and member development" (ibid.), membership appears to be in the interests of local government.

Electoral reform has been contrasted with state restructuring in New Zealand in that rather than being a political initiative, it was a radical change generated by the polity and resented by political and business leaders (Mulgan, 1999). Mulgan goes so far as to say "[i]f anything, electoral change was a reaction against state restructuring, at least against the blitzkrieg methods by which such restructuring was brought about. Though radical in its substance, electoral reform was essentially reactionary in intention, an attempted counter-revolution against the policy-making elites" (Mulgan, 1999:2). Giving the interests of New Zealand's two major political parties in maintaining the status quo, the rivalry between them which led to the referendum on the electoral system was rather ironic.

Government in New Zealand, based as it is on an import from the United Kingdom is thus what is referred to as a "Westminster" system. Great changes have taken place in the government of New Zealand recently, due primarily to changing the electoral system from a Single Member Plurality (SMP)[11] to a Mixed Member Proportional (MMP) form of representation in 1996 (Gallagher, 1998). A Royal Commission appointed to examine the issue concluded in 1986 that MMP would be a definite improvement over the present SMP system, as the latter gave larger parties disproportionate advantages over smaller parties and minorities which the proportionality of MMP would correct. The commission suggested that a public referendum on electoral reform be held on or before the next national election in 1987 (ibid.:26–27). Neither the Labour government—reelected in 1987—nor the

National government, were in favor of replacing New Zealand's SMP, First-Past-the-Post (FPP) system with a MMP electoral system (ibid.) However, as a result of a series of ill-perceived political promises and blunders in response to the commission's 1986 recommendation for change, the National Government under Jim Bolger held an indicative referendum in 1992, with the promise of a binding referendum should the results indicate a definitive wish on the part of New Zealander's for change (ibid.)

While the turnout for the first non-binding referendum September 19, 1992 was low,[12] almost 85 percent voted in favor of changing the electoral system (ibid.:31). The results of the second question the indicative referendum posed—which of four electoral systems the polity preferred: the Single Transferable Vote (STV); Preferential Voting (PV); Supplementary Member (SM) or Mixed Member Proportional (MMP)—indicated that the polity was overwhelmingly in favor of MMP. At the general election in 1993, the National Party was thus compelled to hold a referendum on New Zealand's electoral system. Voter turnout was high and the results of the general election meant that at the next election in 1996, the reelected National party under Bolger would be competing under the MMP system.

The changes under MMP are significant and depending on the success of adaptation to these changes, the implications are great. Following from the results of the binding referendum in 1993, the Electoral Act of 1993 took effect and in April 1995, the Electoral Commission provided for in the act reduced New Zealand's general electorates by thirty-five—resulting in a total of sixty (Levine and Roberts, 1997a). Moreover, a Maori electoral option, while contested, was also introduced, resulting in an increase in Maori parliamentary seats from four to five (ibid.) As politicians became increasingly aware of what opportunities the changes under MMP could present, small parliamentary parties were formed out of defecting National and Labour party politicians. Other small, extraparliamentary alternative parties also emerged, and by the time the 1996 election was to take place, there were twenty-seven parties registered (Harris, 1997).

The results of the first election under the new MMP system in 1996 demonstrate the changes, and results of future elections may do so even more. For the first time since 1928, no political party gained a parliamentary majority in the 1996 election (Levine and Roberts, 1997b). Besides the formation of a center-right coalition government led by the National Party, Labour Party, and New Zealand First (NZF) respectively, two other small parties, the Association of Consumers and Taxpayers (commonly referred to as ACT New Zealand) and United New Zealand were represented in parliament for the first time. In the former SMP system, those aspiring to elected office joined political parties and ran for office as Members of Parliament (MPs) in their electoral district.

The party that had the most MPs voted into the legislature,[13] or House of Representatives as it is called in New Zealand, would then form the government, with the leader of that party becoming the Prime Minister.

Under the new MMP system, there are "two classes of MP: those representing individual electorates (or 'constituencies'), and those drawn from party lists. Previously all MPs were electorate MPs" (Mulgan, 1997:63). The four basic features of MMP are:

> That each voter has two votes, a party vote and an electorate vote; that the basic principle of MMP is that a party's share of all the seats in Parliament is based on its share of all the party votes cast at the election, provided it gets over the threshold; that in order to be eligible to have a share of seats based on its share of all the party votes, a party must cross the threshold by winning at least 5 per cent of all the party votes or by winning at least one electorate seat; that list seats are used to "top up" electorate seats to reach the total number of seats a party is entitled to in Parliament (Harris, 1997:221).

Representation of women and minorities, as is characteristic in Proportionally Representative (PR) systems, has improved (Rule and Zimmerman, 1994) and has as-yet unrealized potential for greater increase. The National Party Prime Minister Jenny Shipley, elected in 1996 as leader of a center-right coalition government pushed for a return to the FPP system or the similar supplementary member (SM) electoral system. The Governor-General recommended and re-appointed[14] by the current Labour-Alliance coalition elected in November 1999 with Prime Minister Helen Clark at the helm, Sir Michael Hardie Boys, has expressed mixed feelings towards MMP on several occasions (Hardie Boys, 1998). According to the Electoral Act of 1993, a review of the MMP system by a Parliamentary Select Committee is being undertaken. With respect to civil defense in New Zealand, the Director of the Ministry for Emergency Management is of the opinion that MMP has exerted considerable influence. He states that "MMP and the resultant political environment demand even greater consensus, excellence and accountability in the Ministry's decisions and activities. Policy advice is also more contestable" (Norton, 1999).

2.5 CIVIL DEFENSE AND DISASTER PLANNING

New Zealand's civil defence organisation was rudimentarily formed in the 1930s and responded to natural disasters such as volcanic eruptions, earthquakes and severe storms which are not uncommon in New Zealand, as well as to military threats (Minister of Civil Defence and the Historical Branch, Department of Internal Affairs, New Zealand, 1990). Like many other institutions in New Zealand, the Ministry of Civil Defence was modeled in large part on its corollary in Great Britain (ibid.) The Public Safety Conservation Act passed in 1932 enabled the government "to proclaim a state of emergency anywhere in the Dominion at any time when "the public safety or public order is or is likely to be imperilled" (ibid.:3). In the event of an emergency, senior police officials in affected areas were expected to take control of the situation "until central control could be established" (ibid.:3).

Emergency response was however, rather ad hoc prior to the passing of the Local Emergency Powers Act in 1953 and the establishment of the Ministry of Civil Defence in 1959 (ibid.) Prior to that, emergency plans of the central government were contained within an Emergency Precautions Scheme (EPS) which was adopted from the British Government and activated in the mid-1930s in New Zealand. The main concerns of this plan, controlled by the Minister of Internal Affairs with representatives from the Police, Defence, Air, Census, Statistics and Labour departments were identified in 1936 as being "earthquakes, air raids and poison gas attacks" (ibid.:3).

The Local Emergency Powers Act of 1953 contained provisions for natural disasters, but the main concern when the act was drawn up was the nuclear threat (ibid.) This act empowered local authorities to set up committees or posts, at any time, for handling emergencies and the potential injuries and unsettlement resulting from them (ibid.) "Apart from loans which might be provided by the Local Authorities Loans Board, there was to be no financial assistance from the central government" (ibid.:8). The act left the non-emergency establishment of civil defense organizations by the localities as optional, and lacked binding prescriptions for supply requisition, compliance to local directives and compensation of emergency personnel injured in the line of duty (ibid.) In addition, the act's emphasis on local empowerment was rather contradictory to plans of the central government contained within the EPS (ibid.)

Attributable in part to some of these shortcomings, the government drew up "its own plan, Government Action in a Major Emergency (GAME). Approved by Cabinet in 1954, GAME comprised instructions to be followed in an emergency by various government departments" (ibid.:9). According to this plan, it was

> the Post Office's responsibility to inform the government of a disaster. A Cabinet subcommittee and an operational centre would then be set up, and if it was necessary to inspect the scene of the disaster, the Prime Minister or his [sic] nominee would visit the site with a field committee which would include the Commissioner of Police, the Chief of General Staff and the Chief of Air Staff. Once on the scene, the field committee would contact the Cabinet subcommit-

> tee to organise government resources. If Wellington suffered a disaster, an Auckland-based emergency committee of district officers would mobilise help for the capital (ibid.:9).

In 1958 the government of New Zealand produced a white paper entitled the Review of Defence in which the argument was forwarded that civil defense planning should receive high priority (ibid.) This was followed by the intragovernmental agency development of "a broad organisational plan, the essentials of which were published early in 1960 in the pamphlet Civil Defence in New Zealand" (ibid.:10;11). The Local Authorities Emergency Powers Act was then repealed by the Civil Defence Act in 1962.

On February 17 of 1964, the first meeting of the National Civil Defence Committee "marked a change of emphasis away from preparedness for nuclear attack. The substance of the policy shift was confirmed with the appearance in 1965 of the plan Government Action in a Major Disaster (GAMD), which emphasised the priority now being given to meeting potential natural disasters" (ibid.:16;17). Three regional divisions were established and entrusted with the responsibility of coordinating civil defence (ibid.)

Following several major natural disasters, including an earthquake, a severe storm and a landslip, important legislative changes regarding civil defense were introduced in 1979 (ibid.) In this year, a "Civil Defence Amendment Act made each regional unit of local government a Civil Defence Region. The three existing Regional Commissioners were each to be responsible for a number of these regions" (ibid.:29).

In 1983, a new Civil Defence Act was passed in an effort to coordinate the civil defense legislation which had been amended six times from the 1962 act, and in order to "define more clearly the tasks of the new regional and united councils in relation to civil defence" (Minister of Civil Defence and the Historical Branch, Department of Internal Affairs, 1990:30). Moreover, incorporation of recommendations of an inquiry into the events of the 1979 Abbotsford Landslip in civil defence planning was desired (ibid.) While a number of amendments have taken place since 1983, and legislative changes affecting the entire organization of civil defence in New Zealand are pending (Ministry for Emergency Management, 1999), the Civil Defence Act of 1983, referred to following the first amendment in 1988 as "the principal act" (Ministry of Civil Defence, New Zealand, 1988) still forms the foundation of New Zealand's present-day civil defense system (Ministry of Civil Defence/Ministry for Emergency Management, New Zealand, 2000). A civil defence emergency is defined by the 1983 act as:

> A situation (not attributable to an attack by an enemy or to any warlike act) that causes or may cause loss of life or injury or distress or in any way endangers the safety of the public and cannot be dealt with by the Police, the New Zealand Fire Service, or otherwise without the adoption of civil defence measures (Ministry of Civil Defence, New Zealand, 1993:7).

More explicitly, civil defense in general is defined by the same act as:

> The measures necessary or desirable for the safety of the public and which are designed to guard against, prevent, reduce, or overcome the effects or possible effects of any explosion, earthquake, eruption, tsunami, land movement, flood, storm, tornado, serious fire, leakage or spillage of any dangerous gas or substance, or other happening that causes or may cause loss of life or injury or distress to persons or in any way endangers or may endanger the safety of the public in New Zealand; and includes the planning, organisation, co-ordination, and implementation of such measures and the conducting of, and participation in, training for such purposes (Ministry of Civil Defence, New Zealand, 1983:Section 2).

The powers of the Civil Defence Act can effectively, "only be applied if an emergency is in force," of which there are four kinds: National Emergency, National Civil Defence Emergency, Regional Civil Defence Emergency, or Local Civil Defence Emergency (Ministry of Civil Defence/Ministry for Emergency Management, New Zealand, 2000). A National Emergency is an atypical event differentiable from the others in that it involves an armed or warlike attack (ibid.) Civil Defence Emergencies at either the national, regional or local level are only called when a situation is determined—as indicated above—as entailing possible mortality, injury, or distress and when such situations are deemed as unmanageable without invoking civil defense structures (ibid.) Procedures invoked during Civil Defence Emergencies are discussed in greater detail following a brief discussion of recent significant legislative changes affecting New Zealand's civil defense system.

In 1989, the Civil Defence Act was amended in response to the dramatic restructuring of local government in New Zealand (Ministry of Civil Defence and the Historical Branch, Department of Internal Affairs, New Zealand, 1990). "The Civil Defence Amendment Act 1989 granted regions the power to contract out civil defence services across boundaries, and to appoint subregional controllers. Districts were given authority to appoint local co-ordinators" (ibid.:45). Analysts within the field point out that the recent reforms within civil defense in New Zealand "need to be viewed within the context of wider social and economic reforms that have taken place in New Zealand since the early 1980s" (Britton and Clark, 1999:1). In 1992 a Department of Internal Affairs review of New Zealand's civil defense system "linked social and economic changes and public sector reform that had occurred since the passing of the Civil Defence Act 1983 with the nation's capacity to respond to a national natural disaster" (ibid.:3). The governmental review stated that the broad reform in New Zealand had "dislocated much of the current Act from modern realities" (Civil Defence Review Panel, cited in Britton and Clark, 1999:3) and that, presented with a major civil emergency on a national scale, the present structures would prove inadequate for handling such a situation (Britton and Clark, 1999).

A year after the governmental review, a National Civil Defence Plan for Disaster Recovery was approved by the Minister of Civil Defence (Ministry of Civil Defence, New Zealand, 1993). The plan stresses the responsibility of local authorities in responding to emergencies that occur in their area and jurisdiction explicitly:

> Local authorities take full first-line responsibility for dealing with the impact of a disaster in their geographic and functional areas of responsibility. This includes the prior provision of insurance and emergency reserves to cover costs associated with disaster recovery. Central government assistance should only be called on where local resources are inadequate to cope with the scope of the disaster or are unable to cope within a reasonable time frame (ibid.:7).

The plan further states that the "aim of any central government assistance is to provide the minimum level of assistance required to restore to the community the capacity for self help, with solutions which provide for the most appropriate long term solution. This does not imply an obligation to restore to previous levels if that is not sustainable in the longer term" (ibid.:3). The circumstances under which the central government will provide assistance to localities are also explicit:

> Central government recovery assistance will normally only be provided where there is a statutory requirement for action, or a need to invoke a statute to achieve the ends desired from the recovery process; or recovery procedures cannot be carried out without central government assistance; or central government assistance will aid the co-ordination of the recovery process to a significant extent; or there are advantages of economies of scale (ibid.:3).

Recently, the central government has clarified the terms and designated the process under which localities may receive financial compensation in a document entitled "Preparing Civil Defence Emergency Expenditure Response Claims: Guidelines for Local Government" (Ministry for Emergency Management, New Zealand, 2000). Local authorities are expected to shoulder the brunt of costs incurred during emergencies and depending on the wealth of a community and amount of money expended in response to an emergency, a locality may indeed find itself responsible for the entire financial burden outside of evacuation costs.[15]

In the event of a declaration of a state of civil defence emergency, "the appointed civil defense controller is responsible for all measures taken to respond to and recover from the emergency" (ibid.:2). In a declared emergency of a civil defence, agricultural or medical nature, coordination of "local recovery efforts" is achieved by the decreed appointment of a Disaster Recovery Co-ordinator, as well as a recovery manager appointed at the regional or district level (ibid.:2). In cases of declared emergency, central government assistance is provided following the recommendation of the Co-ordinator of Domestic and External Security (DESC Co-ordinator), normally to the Chairman of the Cabinet External Relations, Defence and Security Committee (ERDSC) (ibid.:2). The permanently employed DESC Co-ordinator, or, depending on the nature of the emergency, the Chief Executive of the Department affected, decide who should be appointed the Disaster Recovery Co-ordinator (ibid.:7). If a declaration is deemed unnecessary, but the situation is still seen as very serious, the DESC Co-ordinator "may recommend to the Chairman of the Cabinet External Relations, Defence and Security Committee that there is a need for formal action to be taken" (ibid.:2).

While localities are expected to assume much responsibility for their area of jurisdiction both in terms of preparedness and recovery, the central government of New Zealand is committed to restoration of facilities which it owns or provides (ibid.:8). This is not usually the case with state-owned enterprises (SOEs), former government departments that have been converted into businesses—many during the wave of privatization in New Zealand in the late 1980s and early 1990s.[16] The Ministry of Civil Defence expressed concern in the late 1980s that some of these important government departments-turned-SOEs might not be able to function properly in times of emergency (Minister of Defence and the Historical Branch, Department of Internal Affairs, New Zealand, 1990:44;45). The Ministry of Civil Defence appealed to the government to make an exception regarding SOEs, classifying them as organizations rather than enterprises. "The Ministry's wish for SOEs to be classified as "organizations" in terms of the Act was not accepted by Cabinet, but there was a compromise whereby SOEs agreed to carry out their normal corporate functions free of charge during any emergency, and to provide other services on a contractual basis" (ibid.:44;45).

In terms of compensation, the National Civil Defence Plan is clear about which organizations are typically excluded from disaster recovery operations. Restoration of what are classified as "trading utilities" are typically not included in post-disaster recovery operations.

> Government assistance is not normally available for SOEs, local authority trading enterprises, airport and port companies and electricity retailers. Such organisations should maintain sufficient insurance cover and emergency reserves to manage the assured risk. If special problems of risk management and hardship can be demonstrated, government assistance as a special policy may be requested (Ministry of Defence, New Zealand, 1993:8).

However, in the event of a declared emergency, the central government will provide the minimum of assistance possible to local authority assets, which are categorized as generally uninsurable, are not otherwise subsidised and are considered essential services of the community. "Generally uninsurable essential services include water, storm water, electrical, sewerage and gas facilities and other structures, such as retaining walls and tunnels, upon which the essential services depend" (ibid.:9).

As previously noted, community self-sufficiency is stressed in the declaration of states of emergency in New Zealand. While the Civil Defence Act applies to emergency response throughout New Zealand, the central government is currently implementing changes and has devised plans for communities to follow in striving towards the goal of self-sufficiency in emergency preparedness, prevention and response. Many of these changes have stemmed from an Emergency Services Review in 1995 whose Cabinet-appointed Task Force "concluded that there were too many levels and players" (Impact, 1998) and from a Minister of Finance–commissioned Review of Disaster Recovery Preparedness conducted in 1996 and 1997, which gave "particular reference to issues pertaining to the private sector" (Britton and Clark, 1999:4). The Emergency Services Review Task Force identified a need for change in New Zealand's approach to civil defense based on a conviction of consensus around the following:

> There were unrealistically high public expectations of assistance that would be provided. There was a reduced capacity of central and local government to respond following public sector reform. There was a need to improve the ability of the emergency services sector to adapt to changing circumstances, learn from overseas experience, and to better coordinate resources (Britton and Clark, 1994:4).

In accordance with the Task Forces' recommendations, including a new, more comprehensive "structure comprising a Ministry with policy, purchase and audit functions and an operational structure to deal with emergency response that would integrate local and central government emergency service providers," government (National Party) policy in the area of civil defense was introduced (ibid.:4). "Risk Management" is one of the framework principles under the new emergency management arrangements (Britton,

1998). This policy included "the creation of a number of Emergency Management Groups (EMGs) with local responsibilities and the formation of a new Ministry of Civil Defence" (ibid.) The new ministry, the Ministry for Emergency Management, came into being on July 1, 1999 (Ministry for Emergency Management, New Zealand, 1999).

The new Ministry for Emergency Management has adopted an "Emergency Management Group Model" with nine basic principles which local authorities throughout the country are expected to employ in their civil defense schemes. According to principle four of this model, declarations of states of emergency should be "made at the most appropriate level of government by elected representatives" (Emergency Management Group, Wellington, 1999:6).[17] These nine principles are to be applied to a "three-tier emergency response structure" in the declaration of states of emergency (ibid.:12). The local structure of a community, "including local authorities, specialised emergency services agencies and 111 response agencies" are viewed as the "first tier" of emergency response in New Zealand (ibid.:13). It is possible on the first tier for "local authorities to declare at a local authority level" (ibid.) This would involve activating an Emergency Management Group on the local level and responding to the situation through appropriate application of relevant legislation and engagement in planning activities. Should an emergency escalate to the point in which normal emergency management capacity at the local level is inadequate, the "second tier" is activated at the regional level. A second-tier emergency at the regional level is declared by the Chairperson who is the most senior elected official of the regional council (Henare in interview). The second tier involves coordination of the Emergency Management Group on the regional level and development of an Emergency Management Plan which will outline the responses to be coordinated by the Emergency Response Co-ordinator (Emergency Management Group, Wellington, 1999). The Emergency Management Group is to be accountable to the Emergency Management Policy Committee, the governance body of the group which is to be comprised of local elected officials from each of the local areas affected by the emergency (ibid.)

The Chair of the Emergency Management Group is to be an elected official as well, while other members of the group should be senior representatives of appropriate response agencies and local government (ibid.) The "third tier" of response is brought into effect when the emergency escalates beyond the combined resources of the first two tiers and national response to the situation at hand is required. An emergency declared at the national level is to be declared by the Minister of Civil Defence (Henare in interview). Response at the national level is in the process of being revised, but at present, the most recently amended Civil Defence Act is applicable. This act states that when an emergency beyond the scope of local government occurs, the role of the Ministry of Civil Defence within the central government, is "to co-ordinate civil defence measures and to assist regional councils and territorial authorities in their civil defence roles" (Ministry of Defence, New Zealand, 1993/05:1) Once an emergency is declared by one of the above officials at any of these tiers, the Controller, as referred to earlier with reference to the Civil Defence Act, becomes the legally responsible authority for exercising emergency powers (Henare, interview). This individual may in fact also be the mayor or CEO (ibid.)

Emergencies in New Zealand have been few in number and rather small in scale despite the country's rather vulnerable geographic location as a set of small islands forming a part of "the string of highly active seismic and volcanic sites along plate boundaries that stretch from New Zealand, through the Aleutians, and down the Americas," a formation otherwise known as the "Circum-Pacific Ring of Fire" (Britton and Clark, 1999:1). New Zealand's particular vulnerability is a result of its position relative "two lithospheric plates, the Australian Plate and the Pacific Plate. These two plates continue to push and slide against each other and, as a consequence, New Zealand is subject to frequent earthquakes and occasional volcanic eruption" (ibid.) The country also falls within a wind belt, is located on the edge of a major cyclone region and is subject to substantial precipitation normally (ibid.) Surprisingly, over the past 150 years, the average death toll from these sources has been under 3 annually (ibid.) Dependency on modern conveniences and reliance on single sources which have enhanced vulnerability in urban areas seem not to have altered the thought basis of self-sufficiency despite the fact that lifestyles have certainly changed. Illusive or not, what has been described as an "it'll be alright" mentality appears to be embedded in New Zealand culture, for good and for ill from the perspective of emergency management.

2.6 THE AUCKLAND REGION AND ITS INFRASTRUCTUE POLICY: KEY ACTORS

The Auckland Region, administered at the highest level by the Auckland Regional Council (ARC), is made up of seven Territorial Local Authorities (TLAs), of which Auckland City is one. Auckland City Council "is a territorial local authority set up under the central government's Local Government Act 1974" (Auckland City Council, 1998b:8). While the city of Wellington, located on the southern tip of the North Island under the jurisdiction of the Wellington Regional Council, is the capital of New Zealand, it is less than half the size of Auckland City, with a population of 157,646 compared to the more northerly city's 361,000 inhabitants (Auckland City Council, 1998a:13). Auckland City "is the largest multi-functional local authority in New Zealand" (Auckland City Council, 1998b:8).

The Auckland Region is New Zealand's "fastest-growing region. It now has a population of more than 1 million people and the figure is set to reach 1.6 million by the year 2026. In recent years, Auckland has been growing by 24,000 people a year and this is placing pressure on Auckland's infrastructure—especially its roading, and water and wastewater services" (Department of Internal Affairs, New Zealand, 1997:iii). Not only has there been a significant degree of concern about overloaded infrastructure in the area, but there have also been critical changes in the administration of the infrastructure in the region as well. Auckland Regional Council (ARC) was restructured by the central government in 1992 as part of local government reform. At that time, the ARC owned the vast majority New Zealand's largest port, the water infrastructure for the region, valuable commercial property, the Auckland bus company and a number of other assets. The council also has had wide responsibilities in regional planning and regulation as well as in shepherding the numerous regional parks in the Auckland region (McKinlay, 1999).

The central government, expressing concerns that the ARC was becoming too big and powerful and that its considerable indebtedness might prove unmanageable, decided at this time that the ARC would be basically restricted to regulatory and planning activities in addition to the continued control of the parks system (ibid.) All of the ARC's assets and liabilities—including those infrastructural—were transferred into a new institutional structure referred to as the Auckland Regional Services Trust (ARST), which "was not in fact a trust but a statutory corporation governed by members elected at large across the Auckland region" (ibid.:4). The prognosis of financial advisors in 1992 was that if the trust ever generated a profit, it would be many years before that occurred (ibid.) However, by 1998 the ARST had accumulated a net worth estimated at around $750 million NZD (ibid.)

The issue of what to do with assets became increasingly politically salient on every level of government, particularly in the aftermath of the prolonged Auckland brownout, and a great deal of political wrangling ensued. According to the manner in which the body was designed "any surplus of the ARST was to be paid to a community trust. There was no support for that option. Instead contending parties argued variously for dividing the assets among local authorities, distributing the net worth to ratepayers or residents or applying them to specific cultural or other assets within Auckland" (ibid.) On October 1, 1998 under the Local Government Amendment Act of 1998, the ARST assets—excluding Watercare Services Limited, the company responsible for water management in the Auckland region—were ultimately transferred into a new institution referred to as Infrastructure Auckland (IA) (ibid.) Shares in Watercare Services were distributed to local authorities within the Auckland Region according to demand. The legacy of public/private debate surrounding the former ARST has continued with respect to IA as many are in favour of privatisation of the publicly-owned body.

2.6.1 Electricity Provision in Auckland: the Creation of Mercury Energy

The ownership and operation of Mercury Energy Limited, a company formed from a number of complex institutional and organizational changes, has generated a substantial amount of controversy, particularly following the collapse of their power grid supplying Auckland. Following introduction of the Energy Companies Act in 1992, the Auckland Electric Power Board (AEPB) was corporatized and the new private company Auckland Energy Limited (AEL) was formed, which then changed its name to Mercury Energy Limited (Ministry of Commerce, New Zealand, 1998). The transfer of the assets and liabilities of the AEPB into Mercury Energy Limited was completed by April of 1993 under the terms of what was referred to as "The Establishment Plan" (ibid.) Under this plan, the major shareholder of Mercury Energy Limited was designated as the Auckland Energy Consumer Trust (AECT) and the 300 million shares from the AEPB were vested into it (ibid.) The AECT was then

> given 100% ownership of Mercury which had an authorised capital of $400 million. The Establishment Plan, however, provided that 75% of the authorised capital would be held by the Trust and in due course the remaining 25% (the C shares) would be offered to the public through the Initial Public Offering (IPO) (ibid.:111).

The establishment plan "drafted by Russell McVeagh provides for three types of shares" (New Zealand Alliance Party, 1998). The "'A' shares are 'voting' shares held by the Auckland Consumer Trust. 'B' category shares are no-voting shares also held by the Consumer Trust. 'A' and 'B' shares comprise 75% of the total equity of Mercury. However that 75% shareholding entitles the consumer trust to only four out of nine directors" (ibid.) Controversy surrounds the arrangements governing the third type of shares, the 'C' shares. During the power outage, an opposition party drew attention to Mercury's unconventional structure:

> The 'C' shares "are voting shares which are supposed to be sold to the public. These shares are held by six Russel McVeigh lawyers. The intention was for the

> shares to be sold to the public. Even though the 25% public shareholding hasn't been floated, the right to appoint five Mercury Directors was established immediately. That effectively gave Russell McVeagh's six lawyers the right to appoint five of the nine Mercury directors (ibid.)

Thus the C shares which were intended to be available to public shareholders were very significant as they "carried with them the right to 49.5% of the shareholder vote and the right to appoint five of the nine directors to the Board" (Ministry of Commerce, New Zealand, 1998). The Ministerial Inquiry which was appointed by the Ministry of Commerce to investigate the outage confirmed the issue raised by the opposition party during the crisis. The inquiry states: "C shares were issued to six partners of the law firm Russell McVeagh McKenzie Bartleet and are held in trust at the direction of all Mercury directors" (ibid.) The power outage in Auckland left Mercury Energy and members of their board as well as the government of New Zealand vulnerable to a great deal of criticism as the fact that these C shares were in actuality, never publicly floated, was exposed (ibid.) In the aftermath of the power crisis, the AECT attributed the lack of accountability to Mercury, which allowed such an organizational structure to persist to the special arrangements under which Mercury Energy was initially formed (ibid.) To the post-crisis inquiry, the AECT "submitted that the governance structure for the company is unsatisfactory. Despite the Trust being the majority shareholder, it is unable to appoint all the Directors and has no policy formulation role in Mercury" (ibid.:110).

The Establishment Plan under which Mercury Energy was created was significant in that it allowed deviations from the form of public enterprise corporatization favored by the New Zealand government which ostensibly would have enhanced accountability. What has been indicated as an important and purposeful deviation from the Energy Companies Act resulting in reduced accountability was Mercury Energy's not being required under the specially drafted Establishment Plan to publish a Statement of Corporate Intent (SCI).

> A conscious decision was made during the establishment plan process to avoid the application to Mercury of the accountability provisions of the Energy Companies Act 1992. As a result, Mercury was not required to have a statement of corporate intent (ibid.:111).

The Initial Public Offering (IPO) of shares referred to earlier was intended to provide the basis for corporate accountability by allowing for significant public input in the operations of the company, but in the Establishment Plan which formed Mercury Energy, there was no time limit specified for when the IPO must be accomplished (ibid.) According to the AECT, "[t]he failure to implement an IPO to give such accountability, together with the absence of a statement of corporate intent, has resulted, in the opinion of the Trust, in a flawed system of corporate governance and limited accountability" (ibid.:112). While in their submission to the inquiry, the Auckland City Council deemed the AECT a shareholder incapable of "aggressively" managing Mercury's operations as they deemed was required, the AECT trustees defended their efforts to this end, citing for example, a 1994 letter requesting "a mix of financial and economic information about the company [which] went unanswered" (ibid.:112).

As stated, the special arrangement provided in the establishment of Mercury Energy allowed the "C" shares which should have been made public in the IPO to be issued instead to an equal number of partners of the large New Zealand law firm employed by Mercury Energy, Russell McVeagh McKenzie Bartleet. According to the AECT, the allowance of this structure under the Establishment Plan

> has resulted in a situation where, for all practical purposes, the C class shareholders control the board by being able to appoint a majority of the directors. The five class C directors therefore have two roles, first as directors and secondly, as the majority of directors, the ability to control the destiny of those six shares, which in turn could be utilised to prevent the Trust passing a special resolution. Thus the minority shareholders and the majority of the Board have the ultimate ability to exercise control of Mercury, except in relation to major transactions (ibid.:111).

While Mercury Energy defended existing arrangements for ensuring the accountability of their directors, citing the "provisions of the Companies Act 1993 and the common law which [according to Mercury] provide a substantial level of accountability of any director" (ibid.:114), the inquiry into the power outage concluded that "[t]the governance structure of Mercury Energy is unusual and the lines of accountability are not clear" (ibid.) Moreover, the inquiry found that "[n]either the AECT nor the Mercury Energy Board seriously attempted to initiate actions to fill the vacuum prior to the IPO by developing a Memorandum of Understanding or a Statement of Corporate Intent as a means of agreeing on objectives" (ibid.:115). While members of the inquiry did concede that the AECT had indeed attempted to find out how they could tackle the issue of insufficient accountability at Mercury Energy, they noted that the AECT

> failed to pursue its concerns with the Board directly. These matters have only just been brought to a head after a period of five years. The cable failures have given rise to an improved effort of cooperation between the trust and the Mercury Energy

Board. However, the outcome of this cooperation is yet to be measured for its effectiveness (ibid.:115).

Significant changes have taken place in the ownership of Mercury Energy since the Auckland power outage, related to legislative changes requiring the industry to split their lines and energy retailing business in April of 1999. Mercury Energy sold its retailing business to Mighty River Power and changed the name of their lines business to "Vector." After selling off the retail operations to Mighty River for $50 million NZD, Vector is prospering financially (Reynolds, 16/06/99). Despite the fact that the Report of the Ministerial Inquiry into the Auckland Power Supply Failure concluded that the unsatisfactory (yet originally National Government sanctioned) ownership structure of Mercury Energy should be changed, it appears that little but the name changed. "Vector is owned by the Auckland Energy Consumer Trust, but its unusual corporate structure means the trust does not control the company's boardroom" (Reynolds, 16/06/99).

2.6.2 The Lifelines Project

The establishment of Lifelines Projects—projects which identify and link essential infrastructure of communities to organizations which administer and ensure their functioning—are becoming increasingly common initiatives of government and civil defense agencies. The Federal Emergency Management Agency (FEMA) in the United States sees the informed partnerships in Lifelines Projects as integral parts of risk management necessary for building up communities which are "disaster resistant" and is assertively promoting their establishment throughout the US (FEMA, 1999). Similarly-inspired initiatives are common in the emergency management communities of many other nations, including New Zealand.

The Auckland Regional Council (ARC), inspired by successful Lifelines Projects in the cities of Wellington and Christchurch, initiated the establishment of the Auckland Engineering Lifelines Project (AELP) in late 1995 (Wilkie, 1997:v). The AELP produced its first report in July of 1997, which stated the primary purpose of the project and its publications as being "to reduce the impact of all known hazards on the Lifeline services of the Auckland Region" (ibid.) The AELP adopted "an 'All Hazards' approach" whereby a broad possibility of risks including "Natural hazards (earthquake, volcano, cyclone, etc.), Human hazards (hazardous substance spills, vandalism, etc.) and Biological hazards (algae blooms, disease outbreaks, etc.)" could be planned for (ibid.)

In addition to the ARC and The Carson Group, major sponsors of the AELP included the city councils of Auckland, Manukau, North Shore and Waitakere, the district councils of Rodney and Papakura, the Auckland International Airport Limited, the Ministry of Civil Defence, Transit New Zealand, Telecom New Zealand, Power New Zealand, Water Care Services Limited, Enerco New Zealand Limited, Bell South, and EQC. Other involved organizations included the Chamber of Commerce, Ports of Auckland Ltd., Mercury Energy, ACENZ, the Civil and Resource Engineering Department at the University of Auckland, and Willis Corroon Ltd. Eighteen individuals from the participating organizations were to form the Steering Committee of the AELP with most of the project work "being performed by five task Groups, responsible for Hazard analysis, Transportation, Civil Communication and Energy Services" (ibid.)

This first document—The Stage One Report—was to serve as an introduction and orientation for the members, who would become more active over time, through learning workshops such as the first held at the end of July, 1997. It appears however, that only a tenuous grasp of the Lifelines Project had been reached when the electricity outage occurred in January of 1998. In his report based on observation of the crisis entitled "The Auckland Electricity Supply Disruption 1998: Emergency Management Aspects," the Deputy Director of Victoria State Emergency Services (SES) in Australia, Gareth Davis, directly attributes the slowness of the AELP in getting off the ground to the lack of engagement of some of the partners, citing specifically Mercury Energy.

> New Zealand has been ahead of most in putting lifelines planning on the agenda. The lifelines studies on Wellington and Christchurch are documented proof of that. And yet, in Auckland, where stage one report of the Auckland Engineering Lifelines Project was printed in July 1997 there has been difficult in completing the next phase because of lack of commitment by some of the participants. One of those? Mercury Energy (Davis, 1998a:8).

Despite the fact that the project was criticised for being slower in getting off the ground than it should or could have been, it appears that learning has occurred as a result of the crisis, and increased emphasis has been placed on the role of the AELP. During late autumn of 1998, the AELP conducted a seminar entitled "Risk Management—Looking Forward from the Auckland Power Crisis" from which thirteen accounts or analyses were presented by participating organizations. While the sagacity of establishing lifelines projects as an integral part of emergency management planning schemes of local government is increasingly recognized, what has been pointed out as being somewhat problematic in this endeavor "is the coordination required when significant urban areas transcend local government areas," such as is the case in Auckland (ibid.)

NOTES

1. It is important to mention the Cook Islands as they are economically and militarily dependent on New Zealand. The fifteen Cook Islands were transferred to new Zealand in 1901 by the British protectorate which had assumed them in 1888. Following the 1965 granting of home rule, the Cook Islands have been classified as an associated state linked with New Zealand, but with the option of declaring full independence. New Zealand is also responsible for Niue Island and Tokelau Island. The total population of these islands is about 21,000. Another 52,000 islanders live in New Zealand (see Elazer, 1994).

2. The radical restructuring of New Zealand's public sector was theoretically based on "new institutional economics based on property rights theory, transaction costs analysis, and agency theories as well as public choice theories and a strong commitment to what has been referred to as the 'new public management'" (Spicer, Emanuel, and Powell, 1996:4). For more on public sector reform, see Boston, J.'s, "The Theoretical Underpinnings of Public Sector Restructuring in New Zealand," in Boston, Martin, Pallot, and Walsh, eds., *Reshaping the State: New Zeand's Bureaucratic Revolution*, 1991, Oxford University Press, Auckland. For a broader perspective on reform in New Zealand, see for example, James, C., *New Territory: The Transformation of New Zealand 1984–1992*, 1992 Bridget Williams Books, Wellington; Nagel, J., "Social Choice in a Pluralitarian Democracy: The Politics of Market Liberalization in New Zealand" in the *British Journal of Political Science*, v.28, 1998:223–267; "The Kiwi Experiment," *The Economist*, November 3, 1990:17; "Return to Rogernomics," *The Economist*, March 23, 1991:81; "The Mother of All Reformers," *The Economist*, October 16, 1991:18; "New Zealand's Economy: Rogernomics," *The Economist*, October 16, 1993:98.

3. Accounting and management specialists in New Zealand refer to the reforms as occuring in a stage-wise process involving commercialization, corporatization and privitization (Spicer, Emanuel, and Powell, 1996). Commercialization, according to the authors, "involves the restructuring of government departments and functions so as to introduce accountability and economic efficiency into government commercial activities" (ibid.:1). Following from this is corporatization, "an intermediate level of reform. Through this process the trading activities of government departments or enterprises are transferred to state-owned corporations" (ibid.) Corporatization typically does not entail ownership changes, but structural changes requiring the state to function like other private sector businesses (ibid.) "Privatisation is the highest level of reform. It involves the withdrawal of the state from the ownership of the production of goods and services and typically involves the state selling off assets (such as plant, land or forests) or its equity in corporatised trading entities" (ibid.)

4. Immediately following the 1984 snap election called by the subsequently defeated Muldoon, the newly-elected Labour Party under Lange deemed economic conditions in the country so poor that a 20 percent devaluation of the New Zealand dollar was immediately required, as well as a radical program of economic reform. The Labour Party began their restructuring with a campaign entitled "Opening the Books" where they rather unconventionally published Treasury and Reserve Bank briefings in order to

expose what they considered "very poor management by [National] government policymakers" (Dalziel and Lattimore, 1996:23). The Labour Party's initiation of the radical reforms in which "[n]ew right reforms were initiated by the party of the left" is considered paradoxical but explainable by looking at changes in New Zealand politics in the fifteen-year period preceding the reforms (Nagel, 1998:223). According to Nagel, these include "the rise of multi-dimensional politics, a consequent shift in the class composition of the major parties, the closely related phenomenon of Robert Muldoon's dominance over the National Party, the bias of the electoral system against Labour, and the revolt of the New Zealand Party" (ibid.:231).

5. An important exception was granted in the formation of Mercury Energy Limited, the details of which are discussed further in the section "Emergency Provision in Auckland: the Creation of Mercury Energy Limited."

6. New Zealand's policy of "managed care" under corporate arrangements places pressure on health providers to account for and reduce expenditures, a practice which has, in fact, been linked to a number of mortalities at one of the country's major hospitals and led to an inquiry into the matter (Gillet, 1998).

7. The significance of the virtual disappearance of the Ministry of Works to the lack of engineering expertise available during the Auckland power outage was pointed out in the aftermath of the crisis. The point is further discussed in the section, "Experts and Decision Makers."

8. The Ministry of Energy itself was abolished in December of 1989. At this time, the vast majority of responsibilities of this ministry were assumed by the newly-created Energy and Resources Division of the Ministry of Commerce.

9. The privatization of the electricity industry, followed by its increasing deregulation, allowed for the sale of the ECNZ subsidiary Design Power Ltd. to the New York–based multinational Parsons Brinckerhoff (PB) in May of 1998 (Design Power, 1999).

10. The corporatization of the ESAs was made possible first by the passing of the Electric Power Boards Amendment Act 1990, which revised the terms of the Electric Power Board Act 1925, and two years later, by the Energy Companies Act, 1992. When the amendment took place in 1990, there were 61 ESAs: 38 special purpose local authorities; 21 Municipal Electricity Departments (MEDs) under Territorial Local Authorities (TLAs) and two government-owned authorities, Southland Electric Power Supply and Chatham Islands Electricity System.

11. SMP is a plurality system (as opposed to a majority system or system of proportional representation, or a mixed system where different combinations of the three types of electoral systems are employed). SMP is also referred to as a First-Past-the-Post (FPTP) system. See Blaise and Masicotte in *Comparing Democracies: Elections and Voting in Global Perspective*, edited by LeDuc, Niemi and Norris, 1996, Sage Publications, for a comprehensive description of electoral systems.

12. This can, in large part, be explained by the fact that the referendum did not take place at the same time as a general election. Only 55.2 percent voted in the referendum, compared to an average of 80 percent voter turnout for general elections (ibid.:31).

13. The number of MPs in New Zealand has grown from 76 in 1867 to 80 in 1900; 99 in 1993 and, as a result of the Electoral Act of 1993 in preparation for MMP, to 120.

14. The Governor General was first sworn in under Bolger's National Government in 1996.

15. Two types of claim applications may be made under the system: (A) claims pertaining to evacuations necessary during "the period of a national emergency or a legally declared state of civil emergency;" and (B) other costs incurred during and immediately following an emergency (Ministry for Emergency Management, New Zealand, 2000:3). Qualified type A claims are fully reimbursable whereas qualified type B claims allow reimbursement of two-thirds of monies spent by local authorities "in excess of a sum equivalent to 0.01 percent of the net equalisable rateable capital value [NERCV] of the district or districts" affected (ibid.:4). This means for example, that if a community "valued" (i.e. its NERCV) at $82.5 M spent $22,400.00 during an emergency, 0.01 percent of the community's NERCV, i.e. $8,250.00, would be subtracted from the amount spent—leaving $14,150.00—two-thirds of which would be reimbursed, i.e. $9,433.00 (ibid.) In certain extenuating circumstances, exceptions may be made (ibid.)

16. For more on these changes, see the sections dealing with economic and electricity reform in New Zealand.

17. The complete list of principles is as follows:

1. Comprehensive Emergency Management;
2. All Hazards;
3. Integrated Emergency Management systems;
4. Declarations made at the most appropriate level of government by elected representatives;
5. Routine events and emergencies are best handled at the local level wherever possible;
6. Community responsibility and self-reliance;
7. Individual responsibility and self-reliance;
8. Structures underpinned with appropriate technical information and expertise; and
9. Recognition and involvement of volunteer organisations (ibid.:5–6).

3

The Power Outage Crisis: Key Events and Critical Decisions

3.1 THE CRISIS IN BRIEF

The Auckland power crisis took place under unusually hot and dry summer conditions. All four of the main cables transporting electricity to the Auckland Central Business District (CBD)—the core of the urban hub covering an area of 3.5 kilometers—failed completely, one by one, over less than a one-month period, beginning January 22, 1998. The impact on the city was great and as Auckland is the economic locus for New Zealand, reverberated throughout the small island nation. A normal, consistent supply of power to the Auckland CBD was not restored before late April, approximately two months after the complete failure occurred.

The underground system which delivered electricity to the city of Auckland was a cable network owned and operated by Mercury Energy Ltd. comprised of four main 110 kV cables—two oil and two gas. The No. 1 and No. 2 gas-filled cables—Impregnated Gas Pressure Cables (IP)—had been installed in 1959 and the remaining No. 1 and No 2. oil-filled cables—self-contained oil-filled cables (SCOF cables) connected to the IP cables, had been installed in 1976 (Ministry of Commerce, New Zealand, 1998). Six smaller 22 kV cables, referred to as "inter-tie cables" or "interconnectors" linked the Auckland grid together.[1]

The first of the four main cables to fail was the No. 2 IP gas cable running between the Penrose and Quay substations on Thursday, January 22. Two days prior to this first failure, a contractor who had been working at the Liverpool substation—the substation directly linked to the Roskill substation by the No. 1 and No. 2 SCOF cables—inadvertently cut control cables associated with this 110 kV line. This error resulted in the automatic disconnection of the No. 1 110 kV circuit running between the Roskill and Liverpool substations. The disconnection in turn resulted in an overloading of the parallel No. 2 110 kV oil cable, which also disconnected, resulting in a total loss of power to the connected Victoria, Newton, and Liverpool substations. Workers were able to restore power to all affected areas within about forty minutes through rearrangement of the electrical network, but during the outage, the loading on the 110 kV gas cables between the Penrose and Quay substations increased substantially. The increased load of 53 MVA on the gas cables was not regarded as problematic however, as although it grazed the maximum specified rating on the cables—60 MVA—the increase was still deemed to be within the cables' continuous rating (Ministry of Civil Defence, New Zealand, 1998).

Two days later, at 12:29 P.M. on January 22, the first cable failed, the 110 kV No. 2 gas cable running between the Penrose and Quay substations, under a load of only 30 MVA. This first cable failure did not result in a loss of power, but it did cause the parallel gas cable on this line, the No. 1 gas cable, to exceed its specified rating by 20 percent for a short time (ibid). Over two weeks went by without further mishap or power loss in the Auckland area, until February 9, when the No.1 gas cable running between the Penrose and Quay substations failed at 6:01 P.M. under a load of 42 MVA. As rearrangements had been made in the power system following the earlier problems in January, only those customers supplied by the Hobson Zone substation suffered temporary power loss for a period of about thirty minutes. All customers in the Auckland CBD were requested at this time however, to reduce their consumption of electricity by 10 percent. Standby generation was introduced in the Penrose-Quay area, and efforts were made by workers to reconfigure and strengthen the system, particularly in the Quay area, which was seen as most vulnerable. This second gas cable failure on February 9 caused an increased loading of the oil cables between Roskill and Liverpool substations such that, while they remained 20 percent below maximum specified limits, their operation increased to 30 percent over what were regarded as normal operating limits.

On February 13, Mercury Energy announced their plans to construct a temporary 22 kV overhead line between the Liverpool and Quay substations, in order to protect against the potential failure of the existing inter-tie cable betwen the two substations. On February 17, another error was made by a contractor who was working at the Liverpool station. The worker accidentally severed some of the control cables at the Liverpool station, leading to further disruptions to the already tenuous power supply. The following day, Mercury Energy publicly conveyed the fragility of the system and appealed to those in the CBD to further reduce electricity consumption.

Despite efforts to secure the system, the No.1 oil-filled cable running between Roskill and Liverpool substations failed under a loading of 51 MVA on Thursday, February 19. This resulted in the heavy loading and subsequent disconnection of the parallel No. 2. oil cable as well, and a total loss of power supply to the Liverpool, Quay, Newton, and Hobson Zone substations. The magnitude of failure was more pronounced due to the network's earlier reconfiguration following the gas cable failures between the Penrose and Quay substations. Although the No. 1 oil-filled cable was the only one of the oil-filled cables which failed, the protection systems covering the No.2 cable did not work, thus resulting in its disconnection too. While workers were able to reconnect the No. 2 oil cable again, it had been stretched to the limit. Prior to the disconnection of it when the No. 1 oil cable failed, it had reached its maximum continuous rating of 70 MVA. When the No. 1 oil cable failed and the No. 2 cable disconnected, the load on the remaining in-service cables rose to a level about 12 percent over what was normal (Ministry of Civil Defence, New Zealand, 1998).

A customer load management program was introduced by Mercury Energy between 8:00 A.M. and 5:00 P.M. the following day geared towards stabilising the remaining in-service oil cables on the Roskill-Liverpool line at the 12-percent-above-normal load at which they were operating. The company also went out with an appeal to the public to continue with power reductions throughout the weekend. Notwithstanding, the No. 2 110 kV oil cable—the fourth and final main cable supplying the Auckland CBD—failed at 6:00 P.M. Friday, February 20 under a load of 61 MVA. Power was completely cut off from the Liverpool, Quay, Newton, and Hobson Zone substations. In response, Mercury Energy established a 22 kV backup line between Kingsland and Liverpool and strove to reconfigure the disabled network, transferring as much load as possible to substations outside of the CBD (ibid.) Following the final cable failure, Mercury telephoned the Auckland police department to inquire as to the possibility of declaring a state of civil emergency, not realizing that this was a matter for the city's civil defense organization.

Fire alarms began going off all over the CBD due to the lack of power, bringing the issue to the attention of the Auckland Fire Department. At 11:00 P.M., Mercury Energy held an emergency meeting at their offices located outside of the affected area, attended by a number of representatives from the Emergency Services (police, fire, and ambulance) and Auckland City Council. At that time, Mercury also began searching for backup generators.

The morning following the failure of the last main cable, Saturday, February 21, Mercury held its first comprehensive emergency response meeting on their premises, to which all stakeholders were invited. At the broadly-attended meeting, discussion focussed on whether or not a state of civil emergency should be called, something which was ultimately decided against as the Emergency Services were deemed capable of coping with the situation and lives were not regarded as being at risk. On Sunday, February 22, the Mayor of Auckland, Les Mills, issued a public statement in which he appealed to the public to stay out of the CBD due to the power outage. The request was widely observed and later the same day Mills announced that a civil state of emergency would not be invoked.

The first week of March, Mercury Energy and a contractor hired by the company, to assist during the crisis, began surveying the railroad lines leading into the CBD for the location of a new overhead power line to feed power into the CBD. Mercury workers also tried, unsuccessfully, to repair the recommissioned No. 1 and No. 2 oil-filled cables between Roskill and Liverpool substations at this time. Toward the end of the first week of March, questions regarding how many had been affected by the power outage, as well as questions relating to the issue of responsibility, were raised in the media and by politicians. An article in the local daily, the *Auckland Herald*, stated that the workplaces of about 63,000 individuals from approximately 10,000 companies, as well as about 4,000 residents, were without power during the outage. On Monday March 9, the Cabinet agreed upon a Ministerial Inquiry by the Ministry of Commerce to look into the technical aspects of the crisis. Upon the request of Ministers of Civil Defence and Energy, the Director of Civil Defence reviewed the merits of declaring a state of emergency on March 11. The Director concluded that a declaration was unecessary, but added the reservation that his "greatest concern is, and always has been, that no one is seen as being in charge at the local level" (ibid.) On March 23, the Minister of Civil Defence wrote to the Prime Minister explaining that a declaration of state of emergency was unecessary during the Auckland power outage as one of the conditions (that local agencies cannot cope with the situation) as specified in the Civil Defence Act 1983 had not been fulfilled.

On March 25, at the request of the Director of Civil Defence, the Secretary of Civil Defence approved the terms of reference for a report, distinct from but concurrent with the Ministry of Commerce inquiry, reviewing the Auckland power crisis. Among other things, the report investigated the issues of whether it had been nessessary to declare a state of civil defense emergency, how effective crisis response by key actors had been, and the adequacy of the public information process. By the second week of April, the construction of the overhead power line into the CBD was finished and put into use. With operation of this line, outages were effectively eliminated in the area and immediate problems with energy supply to the CBD were terminated. On Monday April 27, Mayor Les Mills publicly announced that businesses and residents could return to the CBD. Plans had been devised to ensure that chaos did not erupt as people rushed back into the district. Contrary to the authorities' expectations, the problem that arose was that virtually no one returned to the restored CBD. Upon recognition of the problem, local politicians embarked on conscious endeavours to draw people back to the area. A trust fund was created to finance such efforts, which continued until the end of June.

A technical report prepared by Integral Energy was presented to the Minister of Energy, Max Bradford, on May 8. The Australian company had been contracted to prepare reports regarding the first question of the Ministerial Inquiry: "What caused the power supply, and in particular, the power cables into the CBD of Auckland, to fail?" (Ministry of Commerce, New Zealand, 1998). Integral Energy's conclusions could be summarized as semicritical while not damning of Mercury Energy's network operation and maintenance. The complete Ministerial Inquiry was published in July, 1998, and the overall conclusions of those who led it were similar to those of Integral Energy, yet more critical, stating that while certain practices were within industry norms, others were not, and the crisis could have been prevented with better planning and maintenance (ibid.)

3.2 CRITICAL DECISIONS AND OPPORTUNITIES

3.2.1 The First Cable Fails

The failure of the first cable, the second gas cable of two located at the Penrose-Quay station (No. 2 Penrose-Quay 110 kV IP gas cable), took place Thursday, January 22 at 12:29 P.M. and did not cause a great deal of concern or hardship. Mercury Energy did not issue a press release or inform other actors of the failure at this time. Their "risk management response to the failure was to treat it as routine" (Ministry of Commerce, New Zealand, 1998:74). Nevertheless Mercury Energy has repeatedly insisted that the system was not overloaded. Significant precursive events did occur in which cable loads were, although within designated operating limits, substantially higher than normal.

Two days prior to the first cable failure, a contractor had accidentally clipped controller cables at the Liverpool Station, cables which were attached to the Mt. Roskill station—one of two entry points where TransPower delivers energy generated to Mercury for distribution via their transmission network.[2] The No. 1 110 kV oil cable between Roskill and Liverpool was immediately disconnected as a result. This in turn resulted in the overloading and subsequent disconnection of the parallel cable, the No. 2

Roskill-Liverpool oil-filled cable. Complete power loss to the Victoria, Newton, and Liverpool Zone Substations, normally supplied by Liverpool 110/22 kV Substation occurred as a result. Full restoration of supply was completed within about forty minutes through rearrangement of the electrical network, but during this process the loading of gas cables on the Penrose-Quay 110 kV cables rose to the maximum specified rating (normal operating levels are about 50 percent of the specified rating). The fleeting increase in load was "about 53 MVA, which is within their [the cables'] continuous rating of 60 MVA (although a higher rating would apply for this short term load)" (MoCD, 1998:3).

When the first cable failure occurred on the No. 2 Penrose-Quay 110 kV gas cable on January 22, supply was not lost, but it did result in the parallel Penrose-Quay 1 110 kV IP cable exceeding its specified rating by 20 percent for a short time (it rose to 58 MVA and then was reduced to 41 MVA). However, Mercury Energy treated the failure as routine because it did not seem to have any severe consequences and it was effectively accommodated by their contingency plan. Raymond Cheng, Manager Network Customer Service, divided the crisis into four parts, as there were four main cables on which the CBD in Auckland was dependent upon for electricity.

> We regarded it [the CBD] as a very important area, so we have done enough planning to measure this contingency. So basically what actually happened in the end was ... it's like driving a new car with a spare tire. If the first puncture takes place, you exchange that with a spare tire. If a second puncture takes place, you're beginning to get worried. When the three or four other tires become punctured, you actually have nothing left to deal with. Did we have a contingency plan? Yes, we did. We really did, because you don't need four power cables to supply power into that area. We have built-in contingency. But what actually happened is something nobody can plan for. Otherwise every car would be carrying four spare tires rather than one (Cheng, interview, 11/16/98).

While Mercury had of course not anticipated such an extreme breakdown, the company was aware of the increasing strain on the infrastructure, brought about by the aging cables and the increased reliance on them due to the burgeoning growth of Auckland. The tunnel which the company had begun building several years prior to the crisis can be seen as a form of proactive contingency planning which unfortunately had not been completed in time.

3.2.2 The Second Cable Fails

When the second gas cable (the No. 1 Penrose-Quay 110 kV IP cable) developed a full fault on Monday, February 9, the impact was much greater.

Mercury Energy personnel were considerably more concerned. At this time, large outages occurred in the Central Business District of Auckland. While spokespersons for the company did not explicitly and publicly state that they perceived the problem as a crisis at this time, Mercury's actions and statements indicate that the failure was indeed seen by some personnel as very serious, would take several weeks to resolve, and was something that could easily lead to more problems in the distribution grid. After the second cable failure had occurred, Mercury began issuing press releases on a daily, and sometimes several times daily, basis. These conveyed the sense of severity with which the second cable failure was perceived. It is interesting to note however, a contradiction between the sometimes grave tones of the press releases and the statements of prominent Mercury Energy employees working behind the scenes in the crisis management of the failures.

Mercury's first press release of many during the crisis stated that, "[a] major cable fault to Auckland's Central Business District power supply will require people and businesses in the area to make an immediate and ongoing reduction in power use for at least the next three weeks." It admitted that the problem was serious and included the first estimation of time required to resolve the problem (Mercury Energy Limited, 02/09/98, 12 P.M.). Included in the same release were statements of the General Manager Network[3] at Mercury Energy, Richard Gibbons, indicating Mercury's awareness that the second fault and redistribution of the network in order to bring more power to Auckland's CBD had caused increased strain on the overall system and vulnerability to further, more extensive outages as a result. Gibbons noted, "the redistribution of electricity load has reduced the overall security of the network and another incident or fault could cause power problems in the areas adjacent to the CBD" (Gibbons, cited in Mercury Energy Limited, 02/09/98, 12 P.M.).

If the power consumption in the CBD was not reduced by at least ten percent directly as advised, Gibbons stated in a release that "we [Mercury] may have to introduce rotating power cuts across the CBD area. These would come without warning and would occur up to three times a day for 30 minutes at a time" (ibid.) Mercury's second, third and fourth press releases issued the following day at 11 A.M., 12 P.M., and 4 P.M. respectively, revealed that the previous day's target had been reached and that "[p]ower supply in the CBD area is stable," but the importance of conservation was reiterated in the midday release, particularly during the hours of peak demand in the afternoon (Mercury Energy Limited, 02/10/98: 11 A.M.; 12 P.M.; 4 P.M.). In addition to calling for reduced consumption of electricity in the CBD by those working and/or living there, Mercury Energy endeavoured to reduce load on the system at this time by appealing to the Ports of Auckland and other large consumers of power to rely on generators for production of electricity (Mercury Energy Limited, 02/10/98: 11 A.M.; 12 P.M.).

At a regular monthly combined meeting of the Group Managers and Directors of the Auckland City Council held the day following the second cable fault, the electricity supply problems in the Auckland CBD were prioritized for discussion (Human Resources Group, Auckland City Council, 02/18/98:1). The main focus of concern at this time appears to have been the planning and coordination of conservation measures at the council offices. The city council assigned its Utility Planning Group within the council to coordinate "Council's power reduction in conjunction with Property and other Council Groups," and "communicating the individual power reduction initiatives agreed directly to the Groups" (ibid.)

Twice daily press releases were issued by Mercury on both Wednesday and Thursday in which it was made clear that the system remained operational, but only just. The appeal for power reductions was repeated (Mercury Energy Limited, 02/11/98; 02/12/98). The company also attempted to assure the public that it was doing all they could to deal with the problem, citing their enlistment of specialists working around the clock and their ongoing investigation of options regarding back-up generation (ibid.)

The second of two press releases issued by Mercury Friday, February 13 announced the development of a new power source for the CBD, which according to Mercury would "remove the need for power reductions and stand-by generation in the area earlier than expected" (Mercury Energy Limited, 02/13/98: 4 P.M.). Mercury stated that the new 22 kV line would be constructed above ground, linking the stations of Liverpool and Quay Streets with works beginning on Tuesday, February 17 and hopefully completed on February 27 (ibid.) Thus, the overhead line, what the General Manager Network had referred to a year previously as "an environmental no-no" was approved following the second cable's collapse (Richard Gibbons cited in Mercury Energy Limited, 1997:10). Mercury's February 15 press release stated that "engineers today completed all network realignments to bring extra power into the CBD," realignments which had begun following the second cable failure on February 9 (Mercury Energy Limited, 02/15/98).

These realignments appeared to be holding well following tests the next day, leading Richard Gibbons to sanction increased usage in Monday's late afternoon release. "Mercury Energy General Manager [Network] Richard Gibbons says today's testing of power realignment on the network found there was now room for some extra load" (Mercury Energy Limited, 02/16/98). Gibbons tempered his comments sanctioning slight increases in power usage for air conditioning by stating that "unnecessary lighting and computers" should be turned-off and blinds closed to sun (ibid.) This release also contained the prediction that a "normal level power supply" would return "in about two weeks." Aucklanders responded to Gibbons' comments by turning up their air conditioners. A reading of Mercury's press release issued the following day suggests that Gibbons' sanction was perhaps unwise, or possibly too well-heeded. The delicately-worded release begins as follows:

> Auckland's remaining power supply cable has handled the slight increase in power use as a result of cooler offices in the CBD today. Mercury Energy General Manager Network Richard Gibbons said yesterday's recommendation that air conditioning system temperatures be reduced from 25 degrees to 22 degrees seems to have been greeted with universal action. "Aucklanders have really cooled off today, with power usage throughout the CBD higher than in previous days, which was to be expected. The good news is that power reduction techniques appear to be continuing," says Mr. Gibbons. (Mercury Energy Limited, 02/17/98).

A certain degree of retraction could be read into Gibbons' comments, as the press release continued with appeals for further reductions. Gibbons stated: "[t]hese ongoing reductions are important as our circuits were at full load during much of the day, even with the assistance of CBD generation. Should anything further happen, such as a rise in temperature, we may need to reinstate the call for extra reductions" (Mercury Energy Limited, 02/17/98). The press release issued by Mercury the day before the third cable went called for further reductions in consumption, noting that the system was "really under pressure," but it did not indicate that a failure was anticipated (Mercury Energy Limited, 02/18/98).

Statements of the Manager Network Customer Services at Mercury Energy during the crisis, Raymond Cheng, the senior manager under Gibbons who was responsible for "power restoration" (Lifelines, 1998) and the bulk of the crisis management activity which went on at Mercury Energy during the outage, leave a somewhat different impression than those of his superior.[4] Cheng's comments indicate that at the time the second cable went, Mercury Energy's sense of threat was not as pervasive as Gibbons suggested:

> After the first two cables failed, we were not panicking because our remaining two power cables, they were adequately looking after the supply into the city. When cable number three failed, we really got very worried. At that time, we started compiling a list of people like me (Cheng, interview, 11/16/98).

Despite the oft-reiterated claim that the complete failure was a contingency which was completely unanticipated and could not be planned for and Cheng's statement that staff at Mercury were "not panicking" prior to failure of the third cable, planning for a complete failure did apparently, begin at the company on February 9. According to a Mercury Energy engineer in the Operations department at Mercury Energy, contingency planning at the company for a complete system breakdown began in earnest following the failure of the second cable.

> The contingency planning that we did on February the 9th—we looked at—if for instance there was a case of perfect failure across Mercury's network, what we would actually do, and the planning stage really hit in so we identified world-wide where there were generators and how we could relocate them here using heavy-lift aircraft—Russian aircraft—so we kind of figured the 20th—on the 20th when effectively everything turned to custard—what we did do is then is started getting them in as fast as we possibly could (Mercury Energy Operations engineer, interview).

The SOE Transpower, owner/operator of New Zealand's national electricity grid issued a press release February 11 commending Mercury's efforts and using management of the event to justify changes in the electricity industry. Transpower had issued a "system security notice" following the second cable failure, something that is done when "grid security risk reach[es] a certain level" (Heaps, 1998). Under the revised contract-based arrangements for maintaining power system security in New Zealand, such a notice from Transpower "gives companies the choice of: managing load reductions, paying for higher cost generation or the increasing risk of power cuts" (ibid.) According to Transpower, "[o]n Monday, load reduced, avoiding high costs and power cuts did not happen. The system worked" (ibid.) It was added that, "[t]he present procedures reflect a transition where responsibility and choice for system security is being passed to power companies. The recent events show the demand side is the most appropriate place to make choices between risk and cost" (ibid.) This optimistic asessment was soon obfuscated by the events that followed.

3.2.3 The Third Cable Fails

Failure of the third cable, the No.1 oil cable at the Roskill-Liverpool station on Thursday February 19 at 3:21 P.M. resulted in the tripping of the No. 2 cable as well, and complete power loss for the CBD of Auckland. This general loss of supply was only temporary. Supply was restored as TransPower was able to get the No. 2 cable at Roskill-Liverpool operating again (Ministry of Commerce, New Zealand, 1998:79). Notwithstanding, the No. 1 cable remained inoperable, meaning that three out of the four main cables supplying Auckland were nonfunctional. Thus there was only partial restoration of power to the CBD, maintained via power reductions (ibid.)[5]

When the third cable failed, "[t]he CEO and Board were notified immediately" (ibid.:79). The failure of the third cable was extremely significant as "there were no prepared response plans, whether for risk management, emergencies, or to establish the exit path from a three-cable failure" (ibid.:80). Mercury Energy did not have a contingency plan in place for a third cable failure, much less for a fourth (which broke down the following day). According to the Ministerial Inquiry which investigated the power outage "[a] possible exit strategy from a three-cable failure was later described by Mercury Energy to the [Ministerial] Inquiry, but this strategy was not a plan in existence on 19 or 20 February" (ibid.)

In a press release issued at 8:30 P.M. Thursday, Mercury's network manager stated that "while the exact cause of the fault is not yet known, it was not a result of overloading" (Gibbons cited in Mercury Energy Limited, 02/19/98). Gibbons continued, stating that "[p]ower reductions have been working well and there was no indication that this cable would suffer damage. While it is not based in the CBD and is not part of the network area we've been so concerned about over the past week, it was bringing power into the area and its failure caused tripping of switches which led to the loss of power throughout the CBD" (ibid.) At this point, Gibbons also instructed people working and/or living in the CBD to return to the full (10 percent) conservation measures they had been advised to adopt directly following the failure of the second cable (ibid.) Gibbons assured the public that Mercury would be working on the problem nonstop until it was repaired, but he could not say when that would be precisely (ibid.)

Gibbons' comments in the first Mercury press release issued Friday at 11 A.M. indicate that the instability of Auckland's power supply was well-known at Mercury. The likelihood of two-hour rotating power cuts across the CBD was stated, with Gibbons noting that "the central city's power supply is on a knife's edge, and had already suffered one cut" that same morning (Gibbons cited in Mercury Energy Limited, 02/20/98, 11 A.M.). It was clear that Mercury knew that an overloading situation was taking place, as they announced the cutting of power to certain regions in order to minimise stress on the overall system (ibid.) Gibbons stated "[a]t 9:24 A.M. we cut power supply to the Karangahape Rd area to avoid damage to the cables supplying that area. We're likely to have to do the same for other areas without notice throughout the rest of the day" (ibid.)

Shortly after midday Friday, a cable was accidentally cut by a contractor at a building site in the CBD, resulting in some power loss in the vicinity

(Mercury Energy Limited, 02/20/98, 2:30 P.M.). The press release which followed stated that the power disturbance resulting from the accidental cable cutting was "unrelated to the current problems" (ibid.) This press release, the last issued before the fourth cable faulted early the same evening, concluded with the information that power cuts were "still a possibility this afternoon and could last as long as two hours," and in order to decrease the possibility of their occurrence, "CBD workers should maintain their power reductions, avoid using lifts and ensure computer files are automatically backed-up on a regular basis" (Gibbons cited in Mercury Energy Limited, 02/20/98, 2:30 P.M.).

3.2.4 The Unthinkable Happens

Early the evening of February 20, the worst-case contingency materialised—the fourth and final cable supplying the CBD of Auckland, the No. 2 Roskill-Liverpool oil cable, failed at 5:33 P.M. (Ministry of Commerce, New Zealand, 1998). This was the "impossible incident" that ostensibly could not have been planned for. As this final cable failure occurred late in the day Friday, Mercury did not issue another press release until 10:30 the following morning. People in the Auckland CBD at the time the fourth cable went did not need a press release to tell them that things had gone from bad to worse. In the CBD "[t]he lights had gone out as well as the lifts, air conditioning and most phone and computer systems. Auckland was in a state of shock and crisis" (Auckland City Council, 1998c).

A flurry of activity began at Mercury Energy and throughout the Auckland area in response to the catastrophic final failure. A degree of confusion initially ensued as Mercury employees, public officials, and civil servants struggled to contact one another and initiate some form of organized response to the calamity. John May, the person responsible for media relations at Auckland City Council, was contacted via telephone by Mercury Energy's CEO Wayne Gilbert at around 6:45 P.M., less than an hour after the fourth cable failed (Hyde, 1998). Gilbert informed May that the situation was dire as all four main cables had failed and what little power being supplied to the CBD was being transmitted via a very insufficient 22 kV fifth cable (ibid.) May was in a predicament when it came to activating the council machinery, as the CEO, Brian Taylor "was overseas and no one could get hold of the mayor. He [Les Mills] was attending a big athletics meeting on the shore" (May, cited in Hyde, 1998:31). May did however, eventually reach acting CEO Bob Wakelin on the telephone and warned him that while it was a little too soon to tell, "we might have some form of civil emergency on our plate" (ibid.)[6]

May then received more calls from personnel at Mercury Energy who were apparently very worried. They stated that "we'd have to evacuate hotels. There were real problems" (ibid.) May responded as follows:

> I said, "Hold tight" because there are specific circumstances for declaring a state of emergency. You only declare a civil emergency if the existing services, like police, fire, and ambulance, can't cope. At that stage there was no indication of that so I spent some time trying to contact the mayor (ibid.)

May had difficulties reaching Mayor Les Mills, but according to May, he did finally reach him at "about 11 o'clock that night and we agreed to organise a meeting the next morning" (ibid.)[7]

While the exact timing of events and precisely what thoughts initiated them is unclear, it would appear that Mercury Energy staff, armed with the cursory information provided by John May regarding civil emergencies, but apparently lacking complete understanding of procedures regarding their invocation, then phoned the Auckland Police Department. According to the Ministry of Civil Defence Review of the outage, Mercury Energy personnel called the Auckland Police Control Room shortly after the final collapse. They were concerned that the outage entailed "a distress issue, and a law and order one" and thought that perhaps a state of emergency should be called (MoCD, 1998:5).[8] Mercury Energy staff evidently did not realize that calling a civil emergency, while decided upon the basis of emergency services' ability to cope with the situation, was in fact "a matter for the City Civil Defence organisation" (ibid.) Whether the response of the operator at the Police Control room who took the Mercury call was affected by their sketchy knowledge of appropriate procedures regarding invocation of civil emergencies or not, the operator reportedly thought that the Mercury employee was a practical joker and thus did not assist (ibid.) In a second attempt, a Police Sergeant acquainted with one of Mercury's staff members was then called from Mercury's offices, "and the Sergeant clarified the situation for senior officers. The situation was outlined to the Police at 10:30 P.M." (ibid.) A meeting was to held that very night at the offices of Mercury Energy (located outside the affected CBD) to discuss the power outage and organize a response to it. It included representatives of all three of the emergency services, i.e. the Police, Fire, and Ambulance Services. While not accorded the ability to declare a state of emergency, Emergency Services, in addition to the Public Health authorities,[9] were the primary actors at this point. As mentioned, it is upon determination of Emergency Services' capability to cope in an abnormal and threatening situation that the Ministry of Civil Defence issues a formal emergency declaration. It was also the Police, Fire, Ambulance, and Public Health authorities who formed the initial problem framing in the crisis. At the Friday night meeting they were informed that the power supply to the 3.5-square-kilometer CBD area had been lost almost completely. They were also informed that this was due to failures in the four main power cables. Mercury further noted that the remaining power that was being supplied through a smaller 22 kV cable could not be guaranteed. Nonetheless, at the "meeting at 11 P.M. Emergency Services indicated that they had the situation in hand" (MoCD, 1998:5), and that therefore, no state of emergency was called for.

The suggestion of an inquiry into the power outage was evidently made late Friday night or early Saturday morning. Mercury's first press release issued at 10:30 A.M., Saturday February 21 was entitled "Mercury Energy Welcomes Review; Accepts Cable Failures are its Responsibility" (Mercury Energy Limited, 02/21/98: 10:30 A.M.). At this time, the CEO of the company, Wayne Gilbert, made his first public statement regarding the crisis. Gilbert stated that Mercury Energy would "welcome any independent review of the failure of four main cables feeding Auckland's Central Business District. I will be suggesting that the President of the Institution of Professional Engineers of New Zealand nominate a person to carry out this review" (Gilbert cited in Mercury Energy Limited, 02/21/98: 10:30 A.M.). The press release concluded with Gilbert's refutation of claims that Mercury Energy had failed to maintain the network in the CBD (Gilbert cited in Mercury Energy Limited, 02/21/98: 10:30 A.M.).

3.2.5 Acute Operational Changes

This decision-making occasion was directly linked to the immediate response to the loss of power in the CBD on Friday, February 20. At the first meeting with Auckland's Emergency Services late Friday night, Mercury informed those present that, first of all, not only had the power supply to the CBD been almost completely cut off by the four cable failures (and they also reported that the fifth cable was grossly inadequate for supporting the CBD), but that there was no short-term solution to the supply problem in the CBD. This initial warning should be juxtaposed with statements that the crisis was only expected to last a week, possibly even two weeks (Les Mills quoted in the Ministry of Civil Defence Inquiry:6; Lindsay in interview, 98/11/17). Compared to the actual duration of the crisis (eight to ten weeks depending on whom you ask) a long-term solution should be seen as one that could fix the problem within two weeks.

The Ambulance Service station in the CBD district was cut off by the cable failures. The station did not have a backup generator but the ambulances can function as independent units without an external power supply. Their dispatch centre is located out in the suburbs and was not affected by the power outage. They also had access to telecommunications in the ambulances, so the incoming calls could be directly dispatched to the ambulances. The St. John's Ambulance Station, which is responsible for providing service for the CBD, is also located virtually across the street from one of the CBD fire stations. This station was also initially cut off but had two smaller generators to keep operations going and could assist the Ambulance Service station, offering them a room, if needed, until Mercury Energy could provide the Ambulance station with a backup generator. The room was prepared at the fire station but was never used, as the Ambulance Service station was able to cope with the situation.

The Fire Service discovered that there was something wrong when, at about 8.30 P.M., fire alarms started going off all over the CBD. Shortly thereafter it was established that these were not actual fires but rather that the automatic alarms were set off by the complete loss of power in many buildings. The Fire Service called a meeting of fire alarm companies to discuss some of the issues related to this. They were informed that the power supply alarms would be out for some time and that the emergency backup power, and reserve batteries for the fire alarms needed to be checked in order to prevent false alarms because of low voltage (Warby, 1998:4).

The safety of the individuals in the CBD quickly became of central concern as the power dwindled. The Fire Service experienced a drastic increase in the number of calls for assistance in the CBD. Operationally they increased the response capability in the CBD by adding two fire trucks and additional staff to run them. The Fire Service prioritized getting people out who were trapped in elevators. They also had to help people out of buildings where the electronic locks had automatically locked the doors due to the power failure. Since the Fire Service did not have keys to let these people out but had to tell the people trapped inside to pull the manual fire alarm the building in order to have the doors open again. There was also a lack of elevator keys and some had to be purchased for the rescue crews. It also proved difficult for the crews to locate the elevators in the elevator shafts. Some of the buildings had more than sixteen elevator shafts and it was difficult to determine which ones people were trapped in. A meeting of the elevator-servicing agents was therefore called to address these issues (Warby, 1998:4).

It also became necessary to check all the buildings and all the elevators thoroughly in order to ensure the public safety. Many of the staircases were completely dark, even though it was light outside, and the rescue services had to buy flashlights and chemical lighting sticks to guide their way as they worked the buildings.[10]

Working the buildings quickly became very physically demanding because there had been a massive heat-up in the buildings, since the air conditioning was not working (Warby, 1998:3). A systemic division of labor was set up between the three emergency services as they helped people who were trapped on high floors or that had tried to walk up the stairs but found that they did not quite have the physical strength to make it. This task is normally performed by just the Ambulance Services. They called the Fire Service for help because of the harsh working conditions in these buildings. Some of the buildings had more than twenty-six floors. With this systematic division of labor, the Ambulance Service was to take the first floors of buildings (since they were deemed the least physically fit) the police would take the next four and the Fire Service took the levels above fifteen (Warby, interview, 98/11/17). Extra Fire Service staff was also called in to work the command unit and from time to time staff was called in for the Operational Energy Management Team (Warby, 1998:3). Because only three fire stations of Auckland's total of thirty were affected by the power failure, sufficient staff was available to cover the additional needs at the other stations (Warby, interview, 98/11/17).

There was also a danger of carbon monoxide poisoning in the buildings that were using portable generators for both the occupants and Fire Service crews. The Fire Service received a number of complaints and concerns about carbon monoxide poisoning from employees of firms where these generators were being used to supply offices with power. Carbon monoxide detectors were purchased and distributed to the fire crews (Warby, 1998:4). There was a worry that people in underground car parks would be subjected to carbon monoxide poisoning, as the ventilation was not working. To prevent this from happening, the car parks were shut down by the City Council, at the advice of the Fire Service.

Strategically the Fire Service set up a special command structure for what was later called "Operation Energy." This type of command structure is often involved in crises (such as large fires) and is part of the Fire Service's General Response Plan. On Saturday February 21, a Special Operations Room was set up at the area headquarters and Chief Fire Officer Ray Warby was assigned to the Operations Commander Post. One person was assigned to Fire Safety, another to Operations and Logistics, and a third was the City Council and Civil Defence Liaison Person (Warby, 1998:3). After having consulted Mercury Energy, the City station and the Fire Service Communications Centre (which is housed in one administrative building) was supplied with a generator so that an uninterrupted power supply could be guaranteed for these key functions (Warby, 1998:4).

It was decided that the emergency risk level in the CBD was to be divided into three alert phases: green, yellow, and red. Each phase had different response plans, determined by the percentage of power Mercury Energy could provide. The green phase corresponded to more than 60 percent of the normal supply requirements to the CBD (New Zealand Fire Services, 1998). The yellow phase corresponded to a 25 to 60 percent supply and this was the phase the crisis was contained in until the power was more permanently restored. The red phase corresponded to a power supply that was less than 25 percent of normal requirements. This phase never materialized although it remained a very substantial threat during the early stages of the crisis since the small 22 kv Penrose cable could have easily been overloaded too.

On Saturday night, February 21, Fire Safety officers visited and inspected all nightclubs, hotels, and entertainment venues in the CBD. The nightclubs and entertainment establishments were prioritized during the night because it was assumed that the owners would want to keep their businesses going at all costs. The CBD is the major entertainment district in the city and large numbers of people tend to come in from the suburbs. The Fire Service and the Police went around to check on the safety of occupants and to give fire safety advice to the owners of different establishments (Warby, 1998:3). Part of the general information drive included putting up fire safety advice and distributing promotional materials in buildings where there were tenants and other occupants. The Operational Crew did this on Sunday, February 22.

As a further step in informing the public of how to deal with this situation, the Fire Service's Fire Safety Demonstrator Caravan was placed downtown on the Aoata Square to answer any questions and to give the public general advice. This caravan is used by the Fire Service for information drives on fire safety in schools and in the communities. The van was well-known to the general public and associated with the Fire Service. It was therefore used to maximize The organization's visibility and presence in the downtown area (Warby, interview, 98/11/17). In addition, a public safety notice was published in the *New Zealand Herald* giving fire safety Advice and a toll-free number for general enquiries during the crisis. Furthermore, the Fire Service chose to take out a paid advertisement in the newspaper to guarantee that the information got out in controlled format and as soon as possible. This communication strategy differed from the other agencies and actors. Mercury Energy and the Auckland City Council were two of the actors that experienced problems trying to get press releases published in the media. When press releases were not submitted as paid ads, the paper would provide the information in the format and at the time that it saw fit. The Fire Service's decision to put out a toll-free number for public inquiries was also an option that the other Emergency Services and authorities chose to forego.

Chief priorities of the Police were to maintain law and order, to protect property, to protect life, and to take care of the traffic situation. Another problem related to the issue of traffic was the fear of a sharp increase in car accidents, since the streetlights at many of the intersections in the CBD were not working. Police officers had to be dispatched on the first night to run the intersections at quite a risk since it was getting dark. Although high levels of incidents were expected, the traffic situation was resolved without a large number of accidents. The Police attribute this to the dedication of the traffic control units and the fact that certain traffic courtesy and caution had developed voluntarily, including reduced speeding.

Another very serious problem that had to be managed early on occurred already on the night of February 20. The tide was coming in and because many of the sewage pumps had been affected by the outage, the sea started pushing up sewage water on to the streets, with such force as to lift manholes. This came rather unexpectedly and the flooding posed a bit of a health hazard until it was managed (Rangihika in interview, 98/11/18). The Police also feared that there would be looting and that other types of crime would rise in the area. This fear was based on the case of the New York blackout in 1977. People raided stores, looted, and destroyed property in the wake of malfunctioning burglar alarms and a swamped police force. This background knowledge, although it was not explicitly referred to as "the New York blackout" to the public, was present in the analogies used by the Police force as they tried to identify some primary concerns.

The Police divided the area into units of one square kilometer where sixteen staff members were put on foot patrol. Traffic Police and recruits were used to cover this activity. The Police also had a helicopter that could survey the area from above and identify traffic congestion, possible crowds, and problem areas in terms of crime. The helicopter was equipped with an infrared scanner that can locate people in the dark. According to the police, this proved to be quite a deterrent to crime.

The Auckland Police were able to cover the extra workload by calling back staff that had been on call or off duty. They coped quite well with the resources available at their station and very little personnel had to be borrowed from other districts. However, media specialists had to be solicited from the National Police Organisation (MoCD, 1998:5). The Police hired two full-time media spokespersons to inform the public of safety issues. Some of the information was coordinated with the Fire Service and the Police, and the Fire Service gave some joint presentations on TV. For the Police, this crisis came to be their most positive media response ever. The Police seldom get to appear on TV to show their work, but this situation proved to be a great public relations opportunity. The very first night of the power outage, the Police actually invited the main TV station to film their strategic meeting at the Police Operations Centre. The turn of events in the CBD, a dropping crime rate and fewer serious injuries triggered very positive media attention for the Police. The CBD District Chief could proudly show journalists asking about looting and raping, that there was nothing inordinate going on downtown (Rangihika, interview, 98/11/18).

The Police's contingency plans were largely based on existing natural disaster plans and operational plans for serious crimes in urban areas. In addition, specialized operational plans for different types of situations were used as building blocks in brainstorming and planning for what could possibly happen in this type of situation. The general awareness of the chaos that developed in the New York blackout in 1977 served an analogy that remained in the minds of the police.

As with the traffic accidents, CBD crime rates dropped rather than rose during the crisis. This was probably caused by the saturation of the area with police officers, the lack of potential victims as people stayed out of the CBD, and the fact that a lot of the bigger companies hired security guards. The most typical crimes that usually plague the CBD are break-ins, car thefts, and burglaries. Vehicles in the streets had become extremely sparse and virtually the only people out were security personnel. On the second day of the crisis, the streetlights had come on again, which greatly helped to reduce risks and the opportunities for crime (Rangihika, interview, 11/18/98).

By the fourteenth day of the crisis the Police shut down their Operations Centre because the situation had stabilized and was slowly getting back on track. The Police continued to monitor the developments in the CBD but they scaled the operation down to a minimum. By this time the key services had obtained generators, and the big businesses had relocated. Some power had also been redirected so that the main intersections in the area were equipped with working traffic lights (ibid.)

Another immediate concern was the state of the medical facilities in the CBD. Two of the city's largest hospitals were situated on the border of the area that was blacked-out. One was the Auckland Hospital (a university hospital with a mortuary) and the other was the Starship Hospital (New Zealand Children's Hospital). These were right on the border of the area affected and were in part supplied with power from lines other than the collapsed ones. The hospitals also had backup generators to keep them going in case of emergency. The hospitals were therefore able to continue running their normal operations but were reinforced with extra backup generators (Mercury Energy Limited Fax, 02/23/98) since the hospital directors expressed concern over any possible further failures.

Dr. Virginia Hope was the Medical Officer on call for Public Health issues when the power crisis broke out. Her superior, the Generic Manager of

Public Health in Auckland, contacted her at her home on Saturday, February 21. Mercury Energy had tried to contact the Public Health authorities on Friday but the phones were down and the emergency call switchboard operators were not aware that this leg of the emergency services could not be reached. Eventually someone reached the Manager of Public Health the next morning and he was able to contact Dr. Hope. She then rounded up the person responsible for food safety and another person on staff, and they went to the morning meeting at Mercury. The Public Health officials realized quite quickly that they needed to reach the people in the CBD a lot sooner than most other authorities in order to give them the information they needed.

Dr. Hope's initial thought was that the main concern was going to be food contamination, due to improperly or inconsistently refrigerated or frozen perishables. Other possible hazards were also on the initial problem list. What they needed to know immediately was if, and how many, residents were in the buildings, and whether there were any problems with the sewage and the water pumps. Dr. Hope and her team met together with persons responsible for environmental health at the City Council after the Mercury meeting on Saturday. They divided up the CBD buildings among them and decided that as many as possible were to be visited the following day. By Sunday the joint team had managed to visit fifty of the four hundred buildings affected. One of the problems the team encountered was that some of the buildings had locked automatically or could only be accessed by the owners and the building inspectors. The team therefore ended up working closely with them. Within the next couple of days the remaining buildings were visited. Fears of flooded basements and sewage overflows proved unfounded, and people had water from time to time as the water tanks were working and filled up during the nights. Potential problems that have figured in other power outages—such as outbreaks of Legionnaire's disease—never developed into actual problems in the Auckland case (Hope, interview, 11/18/98).

The Public Health challenge concerned the hundreds of food outlets in the CBD. Dr. Hope and her team knew that there would be a lack of hot water for cleaning dishes and utensils. They also realized that there may not be enough power for the restaurants to cook at the proper heat and to run their refrigerators and freezers. These places would then not be able to store food safely. A lot of food had probably already gone bad because of this. A related problem to spoiled food would be food disposal. Large amounts of stock would have to be thrown out and the Public Health authorities saw a need to check how the garbage collection was working. Some of the restaurant owners had huge amounts of food stock going bad. Public Health staff were aware that there is always a strong incentive to try and save and use food even though it may not be safe. Many people were trying to save food by putting it in freezers, taking it out as the power disappeared and putting it back in again as the power came on in the evenings. During the first CBD visits on Sunday, the health authorities distributed one-page handouts explaining what the businesses that were going to stay open had to do and how to store their food properly. During the following days the health authorities went back and checked what business owners were actually doing with the food. If they saw that the food was bad, they suggested that the owner get rid of it and if it was still there the second time they came around they helped the owner to dispose of it. Those owners still reluctant to throw out their rotting stock had to be "convinced" or coerced to throw out the bad food by pouring bleach on it. Most people cooperated however, and the City Council and Public Health joint team helped them with the collection of spoiled food (Hope, interview, 11/18/98; Pritchards, interview, 11/18/98). The City Council arranged for extra garbage collection during the first couple of days of the crisis, with trucks attending to the CBD a couple of times each day.

Dr. Hope and her team brought in two extra staff members at the end of the first week. They had enough staff to keep up the supervision of the CBD businesses but the staff was working quite hard and many long hours. Some staff had to be available to cover the night shift, too, because some of the restaurants and bars were only opened at night. The two extra health officials were brought in from outside of Auckland and that made it possible for some of the regular staff to keep their regular hours and their regular tasks. Work started quickly piling up at the Health Authorities, in part since it was the central clearinghouse for all imported foods to New Zealand. The Health Authorities were also responsible for conducting food and disease investigations, and their regular work could not be left unattended for very long. Seized by the intensity of the crisis, most of the regular staff was somewhat reluctant, however, to go back to working behind their desks and would have preferred to stay out on the streets (Pritchards, interview, 11/18/98).

The initial problem framing that took place at this decision-making occasion was one of public safety, law and order, and public health. This changed somewhat over the duration of the crisis, as later decision-making occasions illustrate. At this first stage, however, the pressure was mainly on the Emergency Services. Effective improvisation and good luck prevented any fatalities from occurring. The fact that no one died as a direct consequence of the Auckland power outage has been regarded as one of the most successful (or perhapes fortuitous) elements of the management of this crisis at an operative level.

3.2.6 Developing a Crisis Response Strategy

Concurrently, a more strategic decision occasion was prompted by the need to prepare a comprehensive crisis response that would include both a joint operative response and a strategic response. There was therefore a need to gather all key actors who were either directly affected by the power failure or had to account for the unfolding consequences. Mercury Energy took the first initiative to gather such a response, which had important consequences for the continued management of the crisis.

An initial emergency meeting with Emergency Services and the Auckland City Council had been held at Mercury Energy at 11 P.M. on Friday night, February 20. It was quite clear that a larger meeting was needed which would also include a number of other actors. Such a meeting was scheduled for 10 A.M., the following morning (Saturday, February 21), a few hours after Emergency Services and City Council had met with Mercury Energy. It was Mercury that took charge; they called the meeting, informed the key parties, and hosted it on their premises. It was decided that the meeting would be an open forum. Anyone who wanted to come, and that Mercury judged as being affected by the crisis, was welcome to participate (Rangihika, interview, 11/18/98). Because of this open policy, representatives from business organizations in the CBD, participated along with the responsible authorities and agencies. Of those not invited to the meeting were the media. They were kept outside the building during the meeting. Journalists attempted to get as close as possible and were hanging from whatever they could find outside trying to film through the windows what was going on at the Control Centre (Moore, interview, 11/16/98; Wilson, interview, 11/16/98). Later a special media room was assigned where the media received press releases from Mercury Energy and others attending the meeting.

On Saturday, there was still considerable uncertainty surrounding the situation in the CBD and the reliability of the power supply. The question of whether or not a declaration of emergency was to be made remained on the agenda. Most of the actors, other than Mercury Energy and Emergency Services, had either been informed about the situation by phone the previous night or early that very morning. Most of these people had not been in the CBD when the crisis broke, since they had left work late afternoon on Friday just before the fourth cable broke. Consequently, far from all actors were well informed regarding the situation in the CBD and how serious this situation was going to be. Some of those that had been called the night before had gone down to their offices only to find out that they could not get to the city plans or the necessary statistics stored in computers, and that the offices had no ventilation. These were the kinds of problems discussed at the Saturday morning meeting. A sense of confusion hovered in the minds of many of its participants.

Because the Mercury Energy office was a restricted area, all participants in the meeting had to be issued name tags. This procedure included finding out who was the person assigned to the responsibility of crisis management in each involved party, and it included inviting those representatives to the meeting. Some of the parties, such as the Police, had an eighty-nine-person squad assigned to this task so there was a need to limit the number of people would actually keep attending the meetings. The number of participants was eventually narrowed down to include a few representatives from ten to fifteen different parties.

The Saturday meeting was the first of a long line of briefings that took place at Mercury Energy every day until the crisis was over. First there were three briefings each day but that was reduced to two a day for the duration of the crisis. The participants came from very different backgrounds and a common perspective needed to be fostered. In order to create efficient solutions, Mercury Energy proposed a particular meeting structure that would be applied during the meetings so that the time and resources were used effectively. The problems would have to be clearly identified, resources of the different actors put forth, and potential solutions had to be worked out together in the group so that everyone could follow the proceedings.

Closely linked to these foundational procedures was the idea that the actors would inform each other of what they were doing so that they could air any problems with each other. Perhaps one actor had resources to help another? Within this group of fifteen players it was soon discovered that not everyone was a good speaker and that there was a need to establish business-like procedures. Basic rules of conduct were established. A two-part meeting was also set up. In the first part of the meeting Mercury briefed the participants on the latest developments on the power front and then the various participants, in turn, briefed the Mercury Energy employees and each other of their latest developments. Auckland City Council led the latter part, as the city's governing body. (Cheng, interview, 11/16/98). All decisions were put on an electronic whiteboard from which copies could be distributed in an expedient fashion. This facilitated efforts to keep all actors aware of what had been decided and what was to be done. There was an emphasis on keeping these meetings brief—which was in everybody's interest—but some of the meetings ended up getting rather lengthy anyway.

Mercury Energy also handled a number of other practical issues very smoothly. One of the first things they did was to provide all parties attending the meetings (such as the Mayor) with working space where they could keep contact with their own people and leave messages for their colleagues. All the offices in one area of the Mercury building were made available for the different representatives. This way the actors could keep themselves updated on what was happening at their own workplaces and could quickly update their own staff after the meetings (Cheng, interview, 11/16/98). The people that worked in the operation control room at Mercury, for instance, did not want to go home after awhile. They became so wrapped up in their work, and in all that was at stake, that they simply did not think that they could, or should, leave their positions. These people were under great stress and they played a key role in the coordinating and operative work for the power company during the crisis. The staff at Mercury was put under additional stress as the company was faced with the public's outrage. Mercury worked out strategies for lowering the level of stress and actively monitored their staff as to avoid burnouts.

After the Saturday morning meeting at 12:15 P.M., the Mayor went public with a message that started out saying that "Aucklanders are being asked not to visit Auckland's Central Business District following the power cuts to most of the area" (MoCD, 1998:22). The decision to go out and to urge people to stay out of the CBD area as much as possible was made in consultation with the Police and the Fire Service. This decision was made with the picture in mind of people turning up Monday morning to go to work in the CBD and chaos breaking out. The Police worried about the traffic situation, since most of the traffic lights were out, and they knew most of the facilities were shut down at that time because of the lack of power. The Fire Service was also in favor of keeping as many people out of the district as possible from a public safety perspective. The Fire Service was very busy and did not quite have a full grip on the situation at that point (Warby, interview, 11/17/98). They were also worried about carbon monoxide poisoning in car parks and about people getting trapped in elevators since the power came and went.

The Mayor received a lot of criticism for this message, especially from retailers and other business owners in the CBD. The two business associations in the area were upset with the way the message was presented, since no time limitations nor any explanation as to why people were to stay away were given. Many of the CBD retailers felt that the statement needlessly affected customer behaviour during the first part of the crisis (Heavey, interview, 11/17/98). One of the business associations, Heart of the City, commented on the Mayor's message by saying, "On Day 1 the Mayor states 'stay out of the city' and on Day 4 'I may have overreacted'" (MoCD, 1998:23). Insurance companies with clients in the CBD also experienced complications as a result of the Mayor's message. A representative from the Insurance Council said that some of the retailers on the main streets complained that the message had a devastating effect. As a result "some issues resulting in terms of insurance liability have yet to be finally resolved" (ibid.)

The city's position changed four days later. The City Council explained that they had understood that "the message for the people to stay out of the CBD had not been well received. The mood of the citizens changed by Tuesday, February 24, to see it as an economic problem. The people were encouraged to return to the CBD by Wednesday" (ibid.) The City Council also stated that by this time enough temporary power had been supplied to justify this change of strategy.

In fact, this decision to ask people to stay away from the CBD, may not have been an entirely bad decision from at least two perspectives. From a public safety perspective this was of course a success. The Police and the Fire Service stood firmly by the decision saying that since they were still trying to get a good grip on the situation over the weekend, keeping people out of the CBD was probably the best idea at that time. From a security perspective and with the assessment they could do over the weekend, this was a good way to minimize people being unnecessarily harmed. The Health Authorities had also supported the message as an appropriate one (MoCD, 1998:22). From a crisis management perspective this was also quite a good idea. The "stay away-message" was a good approach to breaking the escalation dynamic of the crisis and for creating some breathing space for the decision makers. Although this may seem to have been rushed and not very well thought out (especially knowing some of its consequences), it was a decision that allowed for the crisis management response team to manage under somewhat lower stress levels. All parties were aware that Monday morning was a sort of deadline, in terms of shaping a response to the situation. If they were not certain of what the situation entailed or how to handle it, bodily harm and possibly chaos could have resulted.

After the first meeting on Saturday, it probably was questionable whether the situation would be under control or that a practical response plan should be ready to be put in place by Monday morning. The decision to prevent as many people as possible turning up in the CBD on Monday morning, was probably a quite effective way of limiting damage, of breaking what may have seemed an inevitable escalation of the situation, and of reducing the stress that the escalation entailed. This was also a good opportunity for the Mayor, who was in a key political position, to demonstrate positive coping skills like resolve, initiative, and command, which are often felt to be important in crisis situations. Symbolic actions are key components to shaping the public's perceptions of civic and political leadership in crisis situations.

3.2.7 To Declare or Not to Declare . . .

Probably the single most important strategic decision occasion during the first three days of the crisis was whether or not to declare a state of emergency. A number of immediate and potential threats to the well-being and the lives of people present in the CBD had been identified by the emergency services. Fear of food poisoning, a lack of fresh water, and carbon monoxide poisoning were just some of these potential threats. A number of potential threats to property had also been identified, such as an expected rise in break-ins and thefts in the area. These were some of the concerns that made the question of whether or not to deploy emergency legislation a prominent one.

The problem of declaring an emergency has been captured by Gareth Davis, the Director of the Victoria Australia's State Emergency Service, and has since been brought up in the Ministry of Civil Defence Inquiry. He reflected:

> We [emergency managers] are continually faced with the dilemma of "when is an emergency an emergency?" As we follow the risk management path we continually debate "how far shall we go?" when discussing threats to the community. If people are not dying or being injured, property not being destroyed or damaged, blood is not flowing, flames not rising, is it our business? I would suggest that the Auckland situation, and others like it, have confirmed that emergency managers can no longer confine themselves to their traditional comfort zones. They cannot isolate themselves from the event, which has none of the normally associated emergency dramatics, but because of economic and sociological effects, is disastrous for the community or the nation (MoCD, 1998:7).

The New Zealand Emergency Management Act prescribes specific conditions under which a declaration of emergency is in order. It also stipulates who is responsible for declaring an emergency. An emergency should be declared on two grounds: when public safety is endangered (i.e. when it may result in loss of life, injury, or distress), and it is found that the Police, Fire Service, or other civil defense measures cannot cope with the situation (ibid.)[11] The Mayor (the chairman) of the City or district that is affected is the official responsible for calling a local state of emergency. If this person is unavailable, the authority is devolved to the Deputy Mayor, and then to the local Controller.

Central government may intervene in a local situation where the Mayor has not declared an emergency but where things are getting out of hand. The Civil Defence Commissioner stationed in the affected area has the power to declare a local or regional emergency on behalf of the central government or the central agency for civil defence if "the situation requires it" (Ministry of Civil Defence/Ministry for Emergency Management, New Zealand, 2000). If a senior police officer considers a declaration of emergency to be required the Police can also request a declaration from the Mayor and, if this is refused, ask the Civil Defence Commissioner to make a declaration (Ministry of Civil Defence/Ministry for Emergency Management, New Zealand, 2000). Had a state of emergency been called during the power crisis, then, according to the existing framework (prior to the crisis entitled the "Emergency Services Coordinating Committee") "the police would have led the emergency services assemblage" (Director cited in MoCD, 1998:13). These were the formal criteria needed to be taken into consideration and the proper procedures required to be followed by the different actors. Aside from these, there were many informal reasons for why different agencies and agents either were in favor of, or not in favor of, declaring a state of emergency.

Mercury Energy had its own interests in having an emergency declared. Mercury had identified possible advantages to such a declaration. It would confirm in people's minds the seriousness of the situation. It would also lead the media's attention away from Mercury towards civil defense and leave the company to deal with the restoration work in peace. It would have also enabled civil defense to put their plans into action in order to help save lives. The company has continuously argued, however, that their attitude toward such a declaration was kept open and was reviewed at least once daily (MoCD, 1998:5).

According to the Civil Defence Act, however, there were insufficient grounds for an emergency to be declared or for economic support to be granted by the Government on the basis of a company not being able to provide any essential services. The Act states that "water, storm water, electrical, sewage and gas facilities and other structures, such as retaining walls and tunnels, upon which the essential services depend" are considered generally as uninsurable essential services (ibid.:9). Thus, the issue of a state of emergency declaration or nondeclaration was—or should have been—largely irrelevant to actors such as Mercury Energy, especially from the perspective of central government compensation. The various public actors each held their own views which are briefly summarised here:

The Fire Service and the Police Department did not think a declaration of emergency was necessary. They felt that the situation was under control and that they could cope with the unusual, but not overwhelming workload. The additional powers that were available under the Civil Defence Act were not deemed necessary. An additional reason for the Fire Service not favoring a declaration was that it would have prevented them from using their own media contacts, which they found to be very valuable (ibid.:6).

The Ambulance Services and the Public Health authorities seemed to have some concerns related to a decision not to declare. The Ambulance Services' main concern was related to the hospitals and the contingency that patients would have to be moved elsewhere. The Ambulance Services' capacity would then have been stretched thin. Such a need did not seem very likely, however, since the hospitals had their own backup generators and were on the border of the affected area and soon got extra generators from Mercury Energy. The Health Authorities would have liked to see more consideration taken of related health issues before the decision was made (ibid.:5).

The Mayor was reported to be strongly against an emergency declaration (ibid.) One account of why he did not favor a declaration was that, in his consultation with Emergency Services, he had been convinced that they were coping adequately with the workload and were not overpowered by the event (Warby, interview, 11/17/98). Another account of why he rejected the idea of a declaration was that he felt that the Auckland emergency services had taken on a coordination role and work pattern, and that he did not see the need to bring in another layer of decision making (Hutchinson, interview 11/18/98; Small, interview, 11/18/98; McKeen, interview, 11/18/98; Wood, interview, 11/18/98). This of course would have happened anyway if the central level were to come in and take over the situation. The Mayor himself said something similar, in that the meeting decided against a declaration because it would have introduced a new dimension to the situation. He also said, "we didn't want to bring in the army or close off the city or forcibly evacuate people" (Les Mills cited in Elias, 02/28/98). The issue of a declaration was revisited in the press and the Mayor was caught saying, interestingly enough in hindsight, "If the prognosis was six weeks or over, 'then that is totally unacceptable' and he would consider calling a state of emergency" (Les Mills cited in MoCD, 1998:6). The community and other actors supported this notion as well.

Mercury Energy initially expected that the crisis would last at least a week, possibly until the following weekend, that is two weeks in total. Businesses in the CBD, in general responded to this expectation by "opting for the dining room table rather than alternative office space to run their businesses from, because they thought it was for a week and they could . . . cope that way" (Lindsay, interview, 11/17/98). The key psychological frame of mind and prognosis therefore seemed to stimulate the idea that this crisis was going to be short so no declaration was necessary. Another factor that may have also influenced the Mayor's position was the timing of the crisis, which occurred just six months before the next election.

The City Council was hesitant. The City Council's entire office building was blacked out and the Council had to spend the first two days reallocating their staff around. They could not access important statistics from their computers concerning the number of people living in the district, maps, and other things needed to handle the situation. From the City Council's perspective the crisis was an electrical and anthropogenic crisis. The City Council is only responsible for handling civil defense in the case of a disaster, which generally refers to a natural disaster of some kind. All of the necessary technical expertise for resolving the crisis was within Mercury Energy itself, hence the Council had none to offer. They did have labor force contractors and these services were offered to Mercury Energy. The Council was not, however, in a position to assess Mercury's technical efforts, whether the company was "right, wrong or indifferent" (Manley, interview, 11/17/98). The Council was confident that, despite whatever had caused the problem, Mercury Energy was giving its best effort to fix it, and the Council was very aware that there was nothing they could contribute should a declaration be made. From a political point of view they were quite happy to leave the management of the crisis in the hands of Mercury Energy. From a legal point of view, the Council did not want to take responsibility away from Mercury Energy. Had the Council decided to support a declaration and take over, they would have also taken on part of the liability (Manley, interview, 11/17/98). The Council was also aware that there were possible financial benefits to declaring an emergency, such as covering the costs of relocating residents, but such considerations did not influence their judgment (MoCD, 1998:5).

City Council's Civil Defence Unit did not really want a declaration either. John Lindsay, the City Council Civil Defence Team Leader, knew that he and his team were not ready for this. He was new on the job and had just started planning and getting his team in order (Lindsay, interview, 11/17/98).

Some of the 'official' reasons for not declaring have been that they thought it might bring undue panic and that the extra powers that a declaration would have brought were not really needed (ibid.)

The Central Government was not in favor of such action, either. The government in office was a coalition government with the parties divided over the Auckland directorate, so there was an interest on behalf of the government to keep Auckland on the right side. Auckland Mayor Les Mill's strongest competition in the upcoming Auckland mayoral election, Christine Fletcher, was a member of the central government at the time of the crisis (Moore, interview, 11/20/98). She had connections with the Prime Minister and on the whole, the government and Cabinet were involved very early on. Declaring a state of emergency would have been a political asset for her in the upcoming election, since it would imply that her opponent was not on top of the situation.

The government took a stand early on in the crisis by deemphasising its own potential engagement in the crisis management effort and steering the discussion of the crisis back to the power company and its legal responsibilities (Lee cited in Hansard, 02/24/98; Tizard cited in Hansard, 02/24/98). The government argued that its role in the crisis was to make sure that the emergency services were coping adequately with the situation and that the outage was handled well from a civil defense perspective. A direct question about the government's role in relation to the Auckland electricity supply failure was answered by the Minister of Energy as follows:

> The Government's primary role is to make sure that the civil defence and public health and safety issues are well managed. Ensuring that the emergency services are in a position to meet any problems until power supplies are restored is the immediate concern of this Government. The Government will back up the work, and has already backed up the work, of Mayor Les Mills and the Auckland City Council. (Bradford, cited in Hansard, 02/24/98)

At the outset of the crisis, one representative stated in Parliament that the situation that had developed was "entirely a contractual matter between the company and its customers" (Fitzsimons cited in Hansard, 02/24/98) and that it had "nothing to do with the Government." (Fitzsimons cited in Hansard, 02/24/98). Other representatives of government, such as the Minister of Energy and the Minister of Police, also expressed their support for and their confidence in the Emergency Services and civil defense offices in Auckland on several occasions in the Parliamentary debate (Bradford cited in Hansard, 02/24/98; Elder cited in Hansard, 02/24/98). Based on these statements and the degree of confidence that the government expressed in the capacity and capability of the emergency services and the civil defense organizations to handle the situation, one can draw the conclusion that they did not think a declaration was neither necessary nor appropriate. The government may also have had longer-term political motives for limiting the central government support and involvement in this case. Prime Minister Jenny Shipley and her government had adopted a tough policy when it came to assisting disaster victims. It had, for example, declined financial support to farmers who had been hard hit by the serious 1997 summer drought. There was therefore a strong incentive for the government to avoid coming out as soft-liners in the Auckland case. This may have influenced the government's view of the crisis, in general. Moreover the government encouraged the customers to sue Mercury Energy arguing that the company, itself, had the responsibility and the finances to cover the costs.

The Ministry of Civil Defence may have had an interest in a declaration, but did not argue against the decision once it was in place. Had a state of emergency been declared, the Ministry of Civil Defence would have played a much greater role in the management of the crisis. This would have given the proponents of the old civil defense system an opportunity to show that the old system worked well and that it was indeed needed. On the other hand, with no declaration, this opportunity was missed. The official statement that there would not be a state of emergency declared came on Sunday, February 22 at 5:15 P.M. It stated: "Auckland's CBD power situation should not become a Civil Defence emergency as people appear to have followed the advice of emergency services" (MoCD, 1998:5). The incident plan that was developed after that included certain triggers that would have led to a reconsideration of the declaration option, something that never actually took place during the rest of the crisis (MoCD, 1998:6).

3.2.8 Media Pressure

The media gathered around the Mercury Energy premises and there was intense pressure on the people working inside the Mercury Energy office, or outside digging up the cables, to provide information to the press. Mercury's office complex is considered a private property with restricted access. In the beginning of the crisis Mercury decided to keep the media off their premises, in part perhaps out of shock at the sudden onslaught of journalists. The media then set up cameras—rigged on poles or hung from lamp posts outside Mercury Energy's windows—in order to obtain inside coverage of the command center (Henare, interview, 11/16/98). In order to remedy what was quickly becoming an uncomfortable working relationship with the media, Mercury Energy set up a special media room approximately a hundred meters from where the command center was located, and staged a daily press briefing at 11:00 A.M. In addition, on a number of occasions, the press was invited to take pictures and walk through the Mercury Energy control center.

Initially, all the participants in the joint crisis group at Mercury made their own individual statements to the press. Despite the fact that they often contained some of the same information, this plurality of "official voices" became quite confusing. In response, Mercury quickly found it desirable to centralize the information that was being released from different sources within the company. As far as Mercury Energy staff was concerned, if any of the company employees were approached with questions from the media they were to refer the inquiries to Mercury's media personnel. The company's executive staff reasoned that if employees could go out and say anything without management support, there could be litigation problems and the company could find itself financially responsible for promises made by its employees (Henare, interview, 11/16/98). The suburban papers, however, reported the contrary. They stated that they had access to Mercury Energy's field staff and that the company had not issued a "no comment" decree (MoCD, 1998:27). Mercury Energy's media spokespeople took part in the daily briefings and were expected by management to articulate the company's present and future policy as well as the progress of the power recovery operation. The rest of Mercury's staff was to get its information from the company's web site. Mercury Energy had the most recent developments posted as media releases on their home page and company staff could just log on to the site to get information on what was being done. Except for the media people and the managers present at the briefings, the staff essentially was given the same information as the media and at the same time.

Issuing joint press releases after every briefing became one of the solutions to the contradictory statements emanating from various actors. In practical terms this was achieved when a routine developed, where at the end of every meeting participants would agree on what to tell the press and how the media should be handled. Mercury Energy's media spokespeople then went around and gathered the information that the different actors wanted to make public and worked it into a joint press release. Subsequently, Mercury Energy staff met the press at the designated media room on Mercury Energy's premises.

Mercury's efforts to consolidate the media releases did not, however, mean that some of the other actors were deterred from putting out their own press releases and holding their own press conferences. The Emergency Services, for example, still issued their own daily press releases, which mainly concerned public safety issues and advice to citizens and businesses.

One of the actors that continued to have a separate media contact outside the joint press releases was the Auckland City Mayor, Les Mills. These separate press conferences did not seem to have affected the credibility of the joint press releases in the eye of the public, but they did, however, create other problems for the Mayor himself. The Mayor was using the office that had been provided for him and his staff at the Mercury office because his own was on the fifteenth floor of a building which was of a blacked out in the CBD. Typically it is the Mayor who plays a key role in declaring a state of emergency and media attention on Les Mills was strong. The Mayor therefore made TV appearances outside Mercury Energy's office and in front of Mercury Energy's logo. This proved to be an awkward political choice because it associated him with Mercury Energy. He was seen as somehow dependent upon the company. As Mercury Energy's credibility with the public eroded when the crisis persisted, so did the Mayor's stature diminish.

In addition to the joint press releases, Mercury Energy continuously issued its own communiqués containing information regarding the technical details of cable reparations, the running of cable tests and the power supply available. The company experienced problems of information distribution. Firstly, no matter how much Mercury offered, the media kept asking for additional, and increasingly more specific information (MoCD, 1998:23). Secondly, it became increasingly evident that the company was grappling with its own persistent technical uncertainty. In line with its overall open media policy, Mercury Energy was very open about the available supply of reserve generator power and how much power had to be saved by the people in order to keep the city lit on an area-rotational basis. The process of testing new cable connections proved to be quite delicate however, and full of setbacks. Optimistic announcements had to be retracted as a consequence. This did not look good to the public. The following examples of contradictory statements were referred to in the Civil Defence Ministerial Inquiry:

> A Mercury e-mail statement of 10:21 on Friday 27 March began, "The power crisis is over—normal power is up and running." This was contradicted only 3 hours later: "The power cut currently affecting about a quarter of the CBD." On Sunday 29th, the Company recommended "those who have generators should keep them as a back up," and on 31 March, "The connection of the overhead link to the Liverpool substation will not be relivened for at least two weeks . . . back-up supply in case of a system failure will not be available as soon as expected" (MoCD, 1998:23).

Mercury Energy management held a firm conviction that people would understand Mercury's predicament and would be cooperative if they were honestly informed and not kept in the dark about the true situation. In the words of Mercury Energy's network manager Raymond Cheng:

> [p]eople don't mind suffering, provided that suffering is shared [and] provided that suffering is informed. Informed meaning, you tell them very honestly what is the total power required, what is the total power [that] is being met, what is amount of power required to be saved by the people and what amount of power can be generated by emergency generation. . . . People will be prepared to co-operate you if you share the problems with them (Cheng, interview, 11/16/98).

The company therefore was firmly committed to keeping an open information policy in order to get the greatest amount of cooperation from the public. Two problems were associated with the estimates that Mercury had made and how they were conveyed to the public. For some reason, despite its obviousness, the fact that restoring power depended on a positive cable test, disappeared from the message as it traveled to the public's ears. There was an expectation buildup every time Mercury talked about preparing a cable to be test-run.

When the power was not restored as a result of failed tests, Mercury Energy's credibility in the eye of the public was damaged. Mercury knew that the power restorations depended solely on the success of the tests, but somehow the public was not aware to the same extent of the conditionality of this statement. Eventually this gap between the sender's intentions and the audience's reception caught Mercury's attention. In order to remedy this, they began stating more and more explicit conditions in conjunction with their power restoration estimates. But this approach came too late. The new statements did not have the clarifying and calming effect Mercury Energy may have been looking for, but only widened the credibility gap that had emerged.

3.2.9 Shortage of Reserve Power and Expertise

The decision to import reserve power was taken almost immediately following the collapse of the fourth cable. The authorities knew that domestic supply of generators and other local power sources would be inadequate. This decision was made during the weekend of February 20–21 at Mercury Energy. The planning behind the decision had begun on February 9. The time that elapsed between the collapse of the fourth cable and the actual importation of reserve power was used to assess the various possibilities and to coordinate delivery.

According to Mercury Energy, these kinds of critical decisions designed only to achieve a quick and effective response and were made without concern for financial implications:

> We considered many, many, many, many options. But in the end, what really counted was the time—minimum time—to be taken to install any of those options; money has never been a problem. I think I'm giving you the right answer here. One, in deciding the emergency options available; one, the availability of such options. Two, money was never a consideration. . . . We were prepared to do whatever we had to do to correct the situation. Number two: what was available on the market at that time; number three: what's practicable to be implemented in the shortest piece of time. Therefore, the emergency line was practical because we happened to have—through emergency consultation with other contractors, consultants [and] availability of resources to build it. Longer term options, better term options such as a power barge from Singapore takes too long. You heard we brought in a ship. . . . only because it happened to be in this area. If it happened [to be] in Sweden, of course, it's no longer an option. Emergency generators—we couldn't bring in too many at once in the beginning—why?—the Gulf War [the second escalation of conflict in the Persian Gulf] was on. . . . Whichever option we have installed in the end [was] based on—not on money—availability, the timeliness and resources available—resources—it

was not money—I must emphasise the manpower, [and] the materials (Cheng, interview, 11/16/98).

Mercury definitely needed to respond to the crisis and provide reserve power immediately. To do otherwise would have been unthinkable given their responsibility as owner of the failed grid and electricity supplier to the community; their very institutional continuity depended on a direct response. That money was, in the words of Cheng, "never a consideration" did not mean that the company did not suffer financially as a result of the crisis. Mercury's 1998 Annual Report stated that the company suffered "a net loss for the 1997–1998 year of $25.3 million, the first negative financial result in the company's history" (Mercury Energy Limited, 1998:4). Still, at the time of the crisis Mercury could not allow money to be seen as a consideration because any indication of cost saving or scrimping in restoration of power to the community would have been perceived by the public and likely emphasised in the media as highly illegitimate behavior.

The emphasis on securing reserve power to the CBD as quickly as possible despite cost was also emphasized by Mercury's Chairman:

> Although there was some loss of revenue from the CBD area, of considerably more significant impact was the cost of restoring power in the fastest possible time and compensating our affected CBD customers. While the final CBD costs are uncertain, we have adopted a conservative approach in assessing them at $128 million (Macaulay, cited in Mercury Energy Limited, 1998:4).

Expertise necessary for providing power to the CBD in the construction of the emergency line and in attempted reparation of the cables was secured, but not without cost to Mercury Energy both in terms of economy and prestige. The company had markedly increased its profit margin since the arrival of its "no-nonsense leader" Wayne Gilbert (ibid). Among other things this had been achieved through major staff reductions "in favour of using outside contractors when required" (*New Zealand Herald*, "In Short," 02/26/98). Aucklanders were aware of the company's success and some of the tactics by which its financial standing was achieved, so despite Mercury's "no-holds barred" expenditures during the crisis, contention bubbled forth—particularly when importation of expertise or "borrowing" staff from competitors became apparent.

Engineers with the expertise necessary to attempt reparation of the damaged and destroyed cables were "unavailable" in New Zealand and had to be flown in from Australia. "With no 110 kv jointers in New Zealand, Energy Australia staff were called in. This kind of work is their 'bread and butter' [said Mercury Energy's] Mr. Gibbons. Local people [could] assist but not actually take charge of the work" (Wallis, 1998:7). The sheer fact that outsider expertise had to be solicited became a bone of contention, which the Minister of Energy defended based on claims of low demand in the parliamentary debate on the power outage held February 24:

> Both the old Auckland Electric Power Board and Mercury Energy have had contracts with the Sydney operators to provide cable-jointing capability on an as-needed basis. For those that believe that we ought to sustain very highly skilled cable jointers in New Zealand, I simply refer to the advice I received that, firstly, there have only been two power failures in these lines in the last two years, and even fewer in the years prior to that, and, secondly, it is impossible to sustain the skills of those people with very little to do. So it was a sensible, technical and commercial decision that the company took (Bradford cited in Hansard, 02/24/98).

Bradford's claims were challenged by several Members of Parliament. Labour MP Pete Hodgson blamed the unavailability of cable jointers in New Zealand on Mercury Energy's downsizing operation and "market solutions," which had entailed a staff reduction of 46 percent over the three to four years preceding the Auckland outage (Hodgson cited in Hansard, 02/24/98). Deputy Leader of the Alliance Party Sandra Lee refuted the claim of contracts with the old Auckland Electric Power Board and Australian cable jointers, stating that the board "actually had its own team of cable jointers on its staff all the time" (Lee cited in Hansard, 02/24/98).

While Deputy Prime Minister Winston Peters (of the New Zealand First party in the coalition government) refuted the responsibility of the current National Party government in the crisis, blaming reforms and restructuring of previous Labour Party regimes for contributing to the current state of affairs, he joined in the criticism of Mercury Energy for their staff reductions (Peters cited in Hansard, 02/24/98). At the onset of the crisis, Mercury was so short-staffed that they not only required expertise from outside the country, but also assistance from other companies within the nation such as Power New Zealand. According to the Deputy PM, Mercury Energy was aware of its image problem early in the crisis: "Mercury Energy actually got Power New Zealand to paint out 'Power New Zealand' on the trucks it was using [to assist Mercury], and the uniforms that its employees are wearing, so that it would not be found out that Mercury Energy was so bereft of staff, expertise and technical assistance" (ibid.) The *New Zealand Herald* also drew a clear connection between Mercury's downsizing and the inability of the company to respond appropriately to the power crisis with sufficient numbers of trained staff. It cited citing the deployment and disguise of rival company Power New Zealand's staff during the outage as an example (*New Zealand Herald*—"In Short," 02/26/98). With regards to expertise, Mercury managers themselves stressed the need for international requisition:

> We have international consultants. We have consultants from the UK, we have consultants from Canada, we have consultants from Australia, and inside New Zealand as well. It's more than one system . . . there's many, many people. Even generators for example, we went international (Cheng, interview, 11/16/98).

The ship "brought in" by Mercury Energy during the crisis for extra power production was a "Trans-Tasman" cargo ship, the Union Rotorua, which Mercury Energy hired during the first week of March (Gardiner, 03/05/98). The boat was anchored in the Auckland harbor and provided extra power, particularly to the company's largest customers (ibid.) The Union Rotorua was capable of producing 12 MW, which added "to the 34 MW of deisel-powered generators already in operation" (ibid.) Ports of Auckland, one of Mercury's biggest clients and a large contributor to the New Zealand economy as a whole was the chief beneficiary of the extra power the cargo ship generated during the crisis. "Power supply from the Union Rotorua berthed at the wharf to Mercury's system was planned for March 9. This would stabilise the total electricity supply to Ports of Auckland and allow all 7 container cranes to operate at normal levels. Cranes had to operate at reduced speed with generator power only" (*New Zealand Dairy Exporter*, 1998:70).

While an invaluable source, the extra power from the Union Rotorua was inadequate for powering the entire CBD. Generators were in great demand throughout the CBD and once domestic supply was exhausted, the company set about aquiring them from other countries. There were not very many generators available for lease within New Zealand. The Electricity Corporation of New Zealand (ECNZ) and the military had a combined total of twenty-four generators which they contributed to the CBD's power needs. On February 24, the Minister of Energy stated in a parliamentary sitting that ECNZ had already "provided four generators to provide back-up assistance for emergency services. This [was] in addition to the 20 medium or large generators that the New Zealand Defence Force [was] able to supply" (Bradford cited in Hansard, 02/24/98).[12]

Due to New Zealand's rather remote island location, heavy-lift aircraft was the only feasible method for transporting generators to the country quickly. Antonov aircraft are the largest in the world and are capable of carrying heavier loads than any other aircraft. Because of the specialized status they enjoy, Antonovs (particularly the An-124) are in demand and are not necessarily readily available for lease:

> The big thing with getting these planes is knowing how to get hold of them. They're Russian military aircraft, and effectively there are three licensed agents around the world where you actually go into a bidding pool, is to, what you have to do is, you have to bid for these planes as to what their availability is. I brought planes out from Milwaukee, I brought planes out from Kiev and I brought the U2 plane[13] from Sidney. I also brought out Russian Illusion aircraft out of Kiev and had them do shuttles across the Tasman—it's slightly smaller than this particular plane (de Abaitua, interview, 11/16/98).

Once shipped, proper installation and operation of diesel-powered generators—particularly over an extended period of time, requires specialized knowledge and expertise as well.

> You also need to ensure that the generator rental company that you're hiring them from has got the support infrastructure. We actually—from MacGregor [in Australia]—which is a Scottish-based company that we used, we brought out four technicians and based them here permanently. So—these were diesel technicians—they kept these things going twenty-four hours a day, seven days—and we never had one major breakdown at all. We also brought in low-voltage cable so we could make the connections through into the switchboards. We also brought out the spare parts for these things as well and got a local infrastructure from Cummings diesel engines here in Auckland to support the service of those machines (de Abaitua, interview, 11/16/98).

Generators—while invaluable and essential resources during power outages—are not the cleanest of machines and when not used properly can be dangerous. Generators run on diesel and must be refuelled regularly. In addition to constituting a fire risk, they produce noxious exhaust and may only be operated in well-ventilated areas. Operation of the machines by persons knowledgeable in their usage is vital for ensuring safety. The power outage in Auckland necessitated the use of generators—particularly at hospitals and other core centers—but fortunately, as the crisis took place during New Zealand's summer, people were not dependent on generator power for heating. While the generators imported by Mercury tended to be professionally deployed and maintained, improvised generation solutions by other parties kept the fire department busy (Warby, interview, 11/17/98).

3.2.10 Who Gets Power?

The supply of available generators within New Zealand had already been exhausted the first couple of days of the power crisis. Mercury's international hunt for reserve power, however, brought in an additional seven generators on February 25, 1998. As generators were being flown in, new problems quickly appeared. The challenge was one of distributive justice. Who was to get the available reserve power? There was a need to create a power priority list, if one did not already exist.

Mercury Energy decided to create a priority list and distributed the generators accordingly, free of charge, for the recipients. The priority list became a controversial issue. Even in retrospect, greatly diverging accounts of the process and the eventual outcome of this distributional decision making exist. Despite considerable disagreement, most actors seemed quite certain on what grounds this decision was to be made and who had the most significant role in preparing the decision. Yet no one seemed quite sure exactly how the reserve power was distributed in the end.

Auckland Regional Council is the government body responsible for resource management. What was at stake in creating a power priority list was within their field of expertise. It was the organization best equipped to answer the question of who should get power first from the supplier when there was a limited supply. The Auckland Regional Council had led a project in collaboration with the Ministry of Civil Defence Auckland, the Lifelines Project, that mapped out the most important infrastructural and organizational units in the Auckland region (Hutchinson, Small, McKeen, and Wood, interview, 11/18/98). From a civil defense perspective, this was exactly what was needed in order to create a useful priority list.

Apparently, the Ministry of Civil Defence made its presence felt at the Mercury meeting and the "Lifelines list" was accepted as the basic solution as a consequence of this pressure (ibid.) The Council said that they sat down and looked at the Lifelines Project and adapted it somewhat by applying common sense to the basic plan. Then they gave the list to Mercury Energy for the company to figure out how they were actually going to implement it. The plan needed some modification because it did not consider all kinds of vital societal functions; it only showed what was needed to guarantee the security of the city's lifelines. For example, a water well that was directly connected to the main water supply pump would show up on the list, but a hospital that was indirectly linked and dependent upon the water supply system would not. These vital functions that had been overlooked in the plan were added and then sent off again to Mercury so they could think of how and where they were actually going place the generators and how the generator service teams were going to be divided (ibid.)

The City Council had also laid claims to the priority list, which was eventually decided by Mercury Energy. The City Council allegedly had a priority list worked out for emergencies based on the Lifelines Project.

Because no state of emergency had been declared, the City Council did not force the implementation of this list but it was presented to Mercury Energy as the priority list that the council would indeed impose if a state of emergency was declared at a later stage. Upon submitting its priority list, City Council expressed its understanding that Mercury naturally had to prioritize among its own customers, but that the list indicated the most important or vulnerable sites in the city.

Another group of actors have been credited with developing an initial priority list; the Ambulance Service together with the Fire Service, the police and Department of Civil Defence Unit in Auckland. This was reported by the Victoria State Emergency Services (VICSES) Deputy Director from Australia who participated as an outside reviewer of the Civil Defence management effort for the Minister of Civil Defence Inquiry (Davis, 1998a).

The priority list presented in the Lifelines Project does not seem to reflect the actual priority list that Mercury decided upon and implemented. The Lifelines Project suggests sites that should be prioritized and protected in the case of an emergency or disaster. The document, however, consisted largely of maps and charts identifying key supply lines, infrastructure and civil defense sites. The document does not contain a short and clear list giving decision makers a quick overview or a condensed checklist to consult, for instance when making a power priority list. The structure of the Lifelines Project report may have had something to do with how much or how little it was actually used when the final priority list was settled. Because the project report contained a rather comprehensive outline and not so easily summarized content, it is possible that whoever was consulting it may not have found it very helpful or that its complicated character created room for Mercury to add and to eliminate some of the recommended protection sites.

Several other reasons for the discrepancy between the list established by the Lifelines Project and the priority list later established have also been suggested. One reason presented was that the Auckland Regional Council, in the midst of this crisis when the issue of a power priority list first came up at the Mercury meetings, simply did not remember that they had conducted this project, and that this had to be brought to their attention by their project collaborators. Forgetting this may simply have been a consequence of intense stress and time pressure, or it may have been that the Regional Council in fact had not forgotten but just did not voice this solution loudly enough.

Finally the Police, the Fire Service's Control Centre, the Auckland Hospital, the Starship Hospital, and Radio New Zealand received generators, along with New Zealand Television (NZTV) and the New Zealand Herald.

The Ports of Auckland also received generator power, as did a number of the downtown hotels. The supply of power to the large downtown hotels proved to be among the most controversial. One part of the controversy was the very high costs incurred by businesses that tried to stay open during the crisis. The excessive rates that the businesses, which did not receive generators from Mercury, were paying for their generator rentals was another bone of contention (Mercury Energy Limited, 02/24/98).

Mercury Energy has argued that the generators were provided to designated Civil Defence sites. Mercury claimed that the hotels were Civil Defence sites and that they were to get power because they could be used as emergency shelters if needed (Mercury Energy Limited, 02/23/98; 02/24/98). Big hotels like the Carlton and the Sheraton received power for this reason. The ambition to open shelters in public facilities apparently developed only some time into the crisis. There was a document entitled, "Draft Welfare Accommodation Plan" forwarded from Auckland's Ministry of Civil Defence office to Auckland Healthcare which dealt with the issue of emergency shelter and included community centers, halls, and charity locales in the plan. This plan was apparently developed several weeks into the crisis and was not finalized until sometime after Friday, March 6 (Ministry of Civil Defence, Auckland, fax 1998).[14]

Another version as to why the big hotels were prioritized for receipt of generator power has been presented. Any industry or business which would create long-term effects should they be closed down (such as the hospitals and the port), were prioritized. This included the big hotels, because shutting them down would have meant sending those people back overseas or to somewhere outside the city of Auckland. That prospect was viewed as having such a detrimental long-term effect on Auckland's image that it was seen as something that must be avoided for the sake of the city's economy (Manley, interview, 11/17/98). Because the threat to public safety was not a major issue in this crisis, it soon developed the characteristics of a primarily economic crisis. One of the primary goals, according to Manley, was to try to ease the economic impact of the power crisis in the CBD. The hotels in fact, were a very important part of the early recovery process.

A third motive might have been straightforward business priorities on the side of Mercury Energy. The large hotels in Auckland were among the first to go to the press and to draw attention to the losses they were suffering because of the power outage. The hotels could easily specify the direct costs because they could point to the percentage of booked reservations and then conclude that the hotels could not host anyone without emergency generators. This has been pinpointed as one of the pressures Mercury was facing when deciding who was going to get the reserve power (Lindsay, interview, 11/17/98).

3.2.11 We Can't Work under These Conditions!

The Auckland City Council and High Court building were both located in the Auckland CBD. Both were left without power when the fourth cable supplying the district failed. In order to continue operations, both soon elected to relocate most of their staff for the duration of the outage. The District Court in contrast was only able to relocate partially during the crisis, and experienced problems in securing generator power and carrying on its functions in the CBD.

The decision to relocate Auckland City Council staff was in line with the decision made by local government—in conjunction with police and emergency services—to urge people working, living, or present in the area to leave the downtown core once it was realized electricity would not be restored to the CBD for quite some time. (May, interview, 11/18/98). The Manager of Public Relations at the Auckland City Council during the power crisis, John May noted that there was

> a collective decision of principally the police, fire, ambulance and ourselves that the city centre should not be the place for people to come to because there was nothing they could do . . . and most firms, which started on the Monday, were ready to shift out. Whether it was the big firms—the lawyers, stockbrokers, people like that, actually shifted out to the suburbs, or in some cases, ran their operations from other cities—Hamilton, Wellington, wherever, so that they were able to offer services to their customers even though they couldn't be physically on the site. We had a problem with this building because it basically didn't have power. So we shifted about 600 people from here out to various satellite offices that we've got. We had to close the art gallery because there was no air conditioning and we couldn't guarantee the safety of exhibits, in terms of the controlled temperatures; we closed for example, our major central library for the same reason—that we were simply unable to operate because there was literally no electricity (May, interview, 11/18/98).

While the vast majority of staff of the city council was, as May pointed out, relocated to suburban offices unaffected by the outage, certain key council personnel remained in the Auckland CBD utilising neighboring facilities. May himself was one of the council employees who remained in the CBD for the duration of the crisis.

> I stayed here, because we actually—we physically were able to use the building next door, which is where our Civil Defence headquarters is. And they were able to maintain a reasonable supply of power. See, one of the problems was that it fluctuated a bit. On occasions there would be some power getting through . . . and some businesses would say "Oh, the power's on!" . . . well it

> was, for about a half an hour and the moment of course the load came on it sort of all had to drop off again. So it wasn't really very satisfactory but we were able to pull in two large generators and eventually we were able to skeleton staff this building and keep our civil defence headquarters next door operational (May, interview, 11/18/98).

With respect to relocating the some three to five thousand CBD residents,[15] Auckland City Council assumed a fair degree of self-sufficiency of the population, as they had with local businesses in the area as well (May, interview, 11/19/98). The Auckland City Council themselves assisted only twenty residents during the crisis to simple accommodations located outside the affected area (ibid.)

The District and High courts of Auckland were also forced to fully or partially relocate. New Zealand's court system is headed by the Court of Appeal, which is the highest court in the country, under which are "three High Courts, 42 District Courts, 14 combined High and District Courts, and a further two District Courts with High Court registries, which means they can carry out the functions of both High and District Courts" (New Zealand Government Online, 1999b). High Courts hear and rule upon "the most serious criminal charges, as well as large or important civil cases, some matrimonial property cases and some appeals from the District Court. The High Court also oversees the power of tribunals" (ibid.) District Courts include "Disputes Tribunals, Family Courts and the Youth Court" (ibid.) Cases which District Courts hear and rule upon include criminal cases, civil cases and smaller-scale financial disputes, as well as cases involving family, marriage, and children (ibid.)

The High Court in Auckland is the largest in New Zealand (Neller, interview, 11/17/98) and the District Court in Auckland is "the busiest court in the country" (Department for Courts, 02/25/98). The High Court in Auckland was fully booked with scheduled cases the week following the collapse of the fourth cable, while the District Court had a number of List Court cases pending from the beginning of the outage and five important jury cases scheduled to begin March 2 (Neller, interview, 11/17/98; Department for Courts, 02/25/98).

At the time the Auckland power crisis struck, the High Court had no "disaster recovery plan" in place, but the Court Manager did have a "civil defence list for all the staff with all their home phone numbers" (Neller, interview, 11/17/98). It was fortuitous for the courts that the full outage occurred late in the day Friday, giving the Court Manager some time to respond, prepare, and invent solutions for accommodating the following week's sittings. She called in eight of her key staff and the Executive Judge to a meeting on Sunday where strategies for coping with the crisis could be implemented for the following week (ibid.)

The relocation of the High Court went smoothly. Once the Court Manager received confirmation from the judges that they wished to continue with their scheduled sittings, she had complete authority to devise relocation solutions. On Monday morning, the manager of the High Court "had the whole court shifted to ten different venues outside the Central Business District and it all went like clockwork" (ibid.) The cooperation and flexibility of the judges and the staff at the High Court were essential in enabling the Court Manager to respond effectively to the predicament and move the court.

> I think the key to it was that the fact that the judges said, "We want to continue sitting, we want to sit on our cases and we will sit anywhere. And that was the most helpful thing they could have said. . . . If I found a venue they would go there. So I didn't have to. . . . I mean I was pretty careful that I was checking the quality of the place and I had to be concerned about the safety of the Judiciary and all that sort of thing, so we checked out personally all the venues on Sunday afternoon (ibid.)

Expediency in acting was essential, as it was evident that many other offices in the CBD were scurrying to relocate their powerless workplaces as well. Despite the fact that the Court Manager and eight of her staff began ringing prospective locales Sunday afternoon "[i]t very quickly became apparent that everybody in the central city was doing the same, because the first places I rang were like conference centres, convention venues, and they were already gone. And I thought, I've got to get ahead off the game here, you know. So I started, I thought, where won't people think of ringing for venues? So I started with the churches, and then I went to the schools, the school board" (ibid.) The High Court manager did briefly contemplate bringing in generator power and remaining at their regular site in the CBD, but elected instead to relocate.

3.2.12 Inquiring Minds Want to Know

The public and other actors representing the public and business community in the Auckland area were dissatisfied and called for inquiries and explanations as to why the crisis could have occurred. Upon failure of the fourth cable, Chairman of the AECT, Stan Lawson declared publicly that while he was not proposing resignations on Mercury's board, he did want a full inquiry into the matter, as well as a guarantee that the Auckland CBD would never again be in such a position of vulnerability (Gardiner, 02/23/98). Dissatisfaction of private citizens was reflected in opinion polls conducted during the crisis and in the lawsuits which began piling up as the outage dragged on. While public opinion called for institutional responses to look into the matter, many private citizens and businesspeople were particularly interested in receiving financial compensation for losses they incurred as a result of the outage.

The appeal for a governmental inquiry into the Auckland power outage came first on Saturday, February 21, the day following the failure of the fourth cable. It was made by Judith Tizard, a Labour Party MP representing Auckland Central (Hansard, 02/24/98). The Cabinet agreed that a governmental inquiry was in order and set about with its arrangements (Gardiner, 02/24/98). The Ministry of Commerce was commissioned for the task. Minister of Energy, Max Bradford, stated at this time that he had been assured of complete cooperation in this aim by both the Chairman of Mercury Energy, Jim Macaulay and by the Mayor of Auckland, Les Mills (ibid.) The Labour opposition's energy spokesperson Pete Hodgson expressed the opinion that "a High Court judge should head the inquiry if the Government wanted the people to accept the integrity of its findings" (Gardiner and Bingham, 03/09/98). Hodgson also called for the inquiry to have terms of reference sufficiently broad to include the controversial "restructuring of the electricity industry and the Government's role in it" (ibid.)

The Ministerial Inquiry into the outage was ultimately comprised of Hugh Rennie, corporate lawyer and Inquiry Chair; Don Sollitt, businessman with electrical engineering expertise; and Dr. Keith Turner, an energy consultant and doctor in electrical engineering.[16] The Ministerial Inquiry was formally agreed to on March 9 (under a directive of the cabinet (CAB (98) M 874)) and was published by the Minister of Commerce in July of 1998.[17] In order to assist the Inquiry in responding to the term of reference which they laid out, "Integral Energy Reports were commissioned at an early stage in the Inquiry's investigations and were further elaborated on and explored during the Public Sittings" (Ministry of Commerce, New Zealand, 1998:182). The Terms of Reference for the Technical Reports which Integral Energy was commissioned to provide covered two aspects: "A review of the distribution planning that gave rise to the installation and ongoing management of the assets that failed; and the causes of the cable failures themselves" (ibid.:177).[18] The Inquiry examined the occurrence of the failures in terms of five dimensions: technical; organizational; operation and maintenance; risk management and contingency planning; and network distribution planning

(Ministry of Commerce, New Zealand, 1998). Selected submissions were allowed on each of the issues examined. Mercury Energy Limited submitted on every one of these issues, detailing their actions and perception of the events.[19]

While the most significant report generated by the crisis was the Ministerial Inquiry into the crisis commissioned by the Ministry of Commerce, other important actors produced documents about the crisis as well, most of them related to their particular organisation's response. The Auckland Regional Council (ARC) and Auckland City Council (ACC) both produced or supported the production of shorter reports. The ARC report was written by Gareth Davies and approached the crisis from an emergency management perspective. The ACC report by John Lindsay focused on hazard and disaster modeling. Increasingly typical of crises—particularly those in industrialized countries and entailing large financial losses—is the interest of the insurance industry in analysing the crisis preparedness, preparation, and response. This is not surprising given insurance companies' commitment to financial reparation of insured clients (cf. Newlove, 1998). The Insurance Council embarked on a report on the crisis during the outage, which it released while it was still underway (Gardiner, 03/31/98). The Insurance Council engaged the expertise of the engineering firm Connell Wagner in Auckland, which reportedly had an outside expert brought in from Britain to try to ascertain how and why the extensive power outage in Auckland occurred (ibid.)

3.2.13 The Cable Tests Keep Failing

The overhead line solution was created in a response to the continuous failures of the CBD cable tests. It had eventually become apparent to the power company and many other actors that the cables would not be back to satisfactory capacity within a foreseeable time. Continuous public revisions of previous statements regarding when the power was expected to be restored, were ruining Mercury Energy's public reputation. There was a need to come up with an additional plan for getting a secure power supply into the CBD. Mercury Energy had to have something else to show the public to prove that they were doing their best to resolve the situation.

The practical solutions to the problems that Mercury Energy worked out during the entire crisis were primarily shaped by two factors: availability and timeliness. Availability of manpower and material were key factors in deciding what solutions would be implemented. The leasing of the Union Rutorua, the major generator ship, was a solution that was adopted mainly because the ship happened to be in the proximity and it could cover part of the need for powering the city. Other solutions, such as importing power lines from Singapore, were discarded as too time consuming (Cheng, interview, 11/16/98). "Money was not an issue" continuously stated the management of Mercury and that seemed to be the case, at least partially, as the company indeed expended considerable amounts of money during the crisis. The total financial hardship put on the company throughout the crisis is virtually impossible to calculate, but it is clear that Mercury Energy spent money whenever it was deemed necessary. Once it was clear that there was not going to be a state of emergency, the financial responsibility clearly fell on Mercury. From that point on the company made a serious effort to minimize its losses in credibility and customer satisfaction by being generous with generators and relief funding.

Mercury Energy sought assistance to build the overhead line solution. An independent contractor, GEC Alsthom, was brought in, which started surveying the railroad lines on Sunday March 1 (Mercury Energy Limited, 03/01/98). Its initial estimate was that it would take seven weeks to build the line. The Mayor found this unacceptable. He felt that public skepticism and anger would erupt if an announcement was made that it would be another eight weeks before the power would be back. The line was then built in five weeks with people working twenty-four hours a day (May, interview, 11/18/98).

The first planning meeting for the overhead line was held on Friday March 6. Closure of the rail corridor was discussed and the Auckland Regional Council provided handouts to the residents indicating which stations would be affected. Newspaper ads and radio announcements were also coordinated by the Regional Council. The message seemed to have reached regular passengers by the following Monday and the closing did not cause many complaints.[20]

The overhead line solution was literally rushed through the bureaucracy. It would normally have taken close to two and a half years to obtain all of the permissions for this kind of operation (May, interview, 11/18/98). One of the main pieces of relevant legislation is the Resource Management Act. This requires that any undertaking that may effect the environment be approved and evaluated by the elected local or regional officials. The company would have had to submit an application that the City Council or the Regional Council would be involved in hearing public submissions on their proposal. The people responsible for making the decisions governed by this act have argued that Mercury Energy never, in fact, would have received the approval to build this overhead line had it not been for the extreme circumstances (Hutchinson, Small, McKeen, and Wood, interview, 11/18/98). In order to get the approval through as quickly as possible the public right to objection was suspended by invoking the emergency powers in the act.

Prior to the disaster, there had been tensions between the Regional Council and their role as the Resource Management Agency and Mercury Energy. A tunnel proposal submitted by Mercury (in order to secure the power supply to the city and to provide the city with new cables) had been delayed a number of times. According to Mercury Energy, the resource management hearings were one of the key factors in delaying the work with the tunnel, which could well have prevented this crisis (Cheng, interview, 11/16/98).

3.2.14 The Lights Are On But Nobody's Home

Finally the power was restored and businesses were allowed to start coming back to the CBD. On Monday April 27, the Mayor welcomed all of the business back into town, but at the same time urged them to return in an orderly fashion. A plan had been produced governing which buildings should be reentered at what time in order to try and manage the inflow of people and to control the usage of power. When "D-day" finally came, nothing happened. Virtually no businesses rushed in to open their offices, and customers were notably absent from the streets of the CBD. This was an entirely different problem than the one the City and Mercury Energy had expected. The City was now facing the prospect of million-dollar vacancies, in skyscrapers and huge office complexes with expensive rents and no tenants returning. The CBD could possibly turn into a ghost area; the pace of the city could slow indefinitely. Mercury was facing a serious loss of capacity in its most profitable consumer area in the city of Auckland.

The fact of the matter was that businesses had adapted to the crisis and had shifted their work in a number of ways which did not immediately lose their attraction or advantages when the power was turned back on.

One basic problem with business not returning to the CBD had to do with the consumer shopping habits. A local business association explained that during the crisis dramatic changes to their consumers' daily routines were necessitated. The businesses that had stayed open during the crisis also had to change their habits. People who used to come into town to do their banking, shopping, or run errands had had to find alternatives during the crisis. They found new bank locations and shopping places. It proved difficult to break these new habits. Similar changes occurred to the businesses who could no longer go to the bank or the post office in the CBD during the crisis. They had to change their work routines to conduct their transactions elsewhere (Heavey, interview, 11/18/98).

Naturally, the decline in business cannot be fully attributed to factors related to the power crisis. It was a hard time for retailers in all of Auckland and the rest of New Zealand in general. Financial turbulence swept the world in 1997–1998 with the Asian crisis, renewed escalation of conflict in the Persian Gulf, and general consumer skepticism prevailed. According to the business representatives, so much bad news in such as short time made it difficult for business to pick up again properly. Hospitality was seen as the sector perhaps hardest hit by the power outage, because a lot of food had perished (Heavey, interview, 11/18/98). Establishments could not provide cold drinks nor cook hot food. In many cases they were not even allowed by the Health Authorities, to use their cutting boards because they were not able to properly wash them in hot water (Hope, interview, 11/18/98; Pritchards, interview, 11/18/98).

After the crisis, one of the first economic efforts was a short-term bounce-back recovery program for small businesses. This effort was put in place not to long before the city discovered that the businesses that had moved out during the crisis were not returning. This was an action mainly aimed at lessening the economic impact of the crisis on businesses unable to move out that had suffered badly because their suppliers and customers had left the district. The Mayor told the Mercury Energy Board some time into the crisis that people were suffering because of lost income. He suggested that Mercury set up a trust fund for the immediate financial needs and for the survival of businesses in the CBD. Mercury Energy agreed and allocated $2 million NZD for the urgent needs of their customers, focusing mainly on the small businesses. The Mayor's chief political opponent, however, was on the Board at Mercury and had no intention of letting the Mayor take credit for this action, and therefore insisted that the two million be split and distributed by the City Council and the Chamber of Commerce. This was also how the money was distributed and it caused a lot of bureaucratic friction in what was hoped to be a smooth and effective relief effort (Manley, interview, 11/17/98).

The City Council was quite a natural choice for this kind of trust, because it had plenty of experience in giving out grants. Since two agencies were to distribute financial assistance to the same power customers intensive coordination was needed to prevent inefficiencies and fraud in the use of the relief funds. Mercury's financial assistance, however, was distributed over four weeks. The City Council contributed with another million and a half for this immediate bounce-back fund.

Claimants would literally line up on the side of the street and if they had a business in the CBD, they would be entitled to receive an amount of money. The criteria for receiving Mercury's financial assistance were that the claimants to have a business and were Mercury Energy customers. This criteria proved too strict and therefore the liquidation of the fund took longer than expected (May, interview, 11/18/98). There were nevertheless, people that did not qualify as business owners or managers but that were hurting badly by the crisis either as a resident or otherwise. These people were able to get some financial assistance too because a generous Japanese gentleman had simply walked into the financial assistance centre with a ten thousand dollar check to be distributed to whoever needed it. The criterion for getting money from this smaller fund, was then that a person simply was in a tough spot and needed some money to tide them over until the crisis was resolved (Manley, interview).

The largest economic contribution to the recovery of the CBD was the $5 million NZD allocated by the Auckland City Council. This effort became significant after it was discovered that businesses and customers that had moved out were not returning at the expected rate. The money had been allocated early on in the crisis but the actual plan, specifying how to spend the money was not fully completed until late in the crisis as part of the recovery phase. The city allocated the $5 million NZD after the mayor had proposed to do so at an emergency Council meeting. Mayor Mills put together this proposal with the help of Clive Manley, who was responsible for the disaster recovery in the City Council. It was Manley's experience that in order to get the most money out of a community for the disaster recovery, the money should be asked for fairly early in the crisis. People are then still uncertain of the long-term impact of the disaster and they still feel a sense of solidarity with the victims, in particular, and with the community, as a whole. The mayor and Manley therefore designed a request for recovery funds from the Council very early on and the mayor presented it during the first week of the crisis. Like in most other budget negotiations, the mayor probably asked for more money than he expected to get. He put forward a suggestion for five million and surprisingly obtained the entire sum (Manley, interview, 11/17/98).

It has been suggested that the politicians on the Council voted for this suggestion without really knowing some of the implications. Following the approval by the Council, the mayor and Manley quickly set up a trust fund and put the majority of the money in that trust. If the mayor and his collaborators would have waited another week the Council probably would not have approved it. Once the money had been placed in the trust, the Council could not pull back the money. The money was then spent in accordance with what the board of the trust deemed appropriate (Manley, interview, 11/17/98).

A focus group was initially summoned to deal with the problems related to the economic recovery of the CBD. These included Mercury Energy, the Mayor, a couple of City Councillors (including Clive Manley), Tourism Auckland, the CBD's two business organizations (K' Road Business Association and Heart of the City), the Retail Merchants Association, the Chamber of Commerce, and representatives from the entertainment complex, Sky City. These parties subsequently became the trustees of the CBD recovery trust (Heavey, interview, 11/18/98).

The trust created an umbrella campaign called 'Good To Be Back' that was to spend the remaining $3.5 million NZD of the Council's money. That campaign included TV advertisements, billboards, and a certain amount of promotion on the east coast of Australia. The message was mainly that "the CBD is up and running," "it is good to be back," and "come out and support the Auckland CBD." There were also a number of specific activities going on under this umbrella campaign of which the "Million Dollar Shop 'til You Drop" was the most important one. That event was arranged so that the people that shopped in the CBD during a certain period of time got a drawing ticket with their receipt. The prize money was altogether one million dollars which could only be spent in the CBD and had to be spent entirely in just one day. This meant a straight cash injection into the CBD. At the same time people were being drawn into the CBD which all businesses benefited from. The "Million Dollar Shop 'til You Drop" event also received ample TV and media coverage, which helped spread the picture of a revived CBD (Heavey, interview, 11/18/98).

A good deal of the money was spent on getting advice on how to set up events big enough to draw people from the outer suburbs. Two firms were hired as consultants in order to arrange campaigns and to generate ideas for how to get businesses back in the CBD. One was an advertising and the other a public relations firm (ibid.) Some of the other activities that were going on under the "Good To Be Back" campaign was a free concert with the pop group Aqua. This drew many people wandering through the CBD after the concert, generating business for the local restaurants and bars. The streets were closed off to traffic, street fairs arranged and many smaller activities were organized (Manley, interview, 11/17/98).

The trust ceased operations at the end of June 1998 and the board was dissolved. By then most, but not all, of the five million had been spent. It proved difficult to spend all of the allocated Council funds in such a short time and a few hundred thousand dollars were therefore returned to the Council (Heavey, interview, 11/18/98).

NOTES

1. See figure depicting Mercury Energy's grid in Appendix.

2. See figure in Appendix.

3. Mercury Energy's organizational structure is such that under the CEO and their personal assistant, there are ten "General Managers" in the following areas of specialization: Network; Marketing; Futures; Generation; Special Projects; Strategic Support; Finance; Customer Service; Wholesale Supply; and Information Services (Ministry of Commerce, New Zealand, 1998:30). Gibbons, as Mercury's General Manager Network at the time was "according to his job description, the person responsible for the ownership, acquisition, development, planning, design, construction and maintenance of Mercury's distribution assets. This includes lines, cables, substations and auxiliary equipment" (Reynolds, 02/23/98:E3). See organizational chart of Mercury Energy in Appendices.

4. The General Manager Network supervised by Richard Gibbons was comprised of six managers in the following areas of specialization: Planning; Customer Service; Projects; Process Improvement; Transport; and Business Development (Ministry of Commerce, New Zealand, 1998:30). Mercury Energy's Manager Network Customer Services at the time of the crisis was Raymond Cheng.

5. There was a fifth cable in the Auckland CBD which should be mentioned even though it was of such limited capacity that it had minimal bearing on the crisis. It was a 22kV cable which ran from Roskill through Kingsland to Liverpool and it functioned in a limited way during the crisis by helping "maintain supply to key institutions in the CBD and to allow systems to recharge overnight" (MoCD, 1998:6).

6. This is a very abridged account of Gilbert's rather unconventional contact with May and May's subsequent contact with acting CEO of the council. This situation and May's difficulties in reaching the acting CEO and Mayor of City Council are discussed in detail in "Crisis Communication and the Mass Media."

7. The central government authorities were notified of the collapse of the fourth cable at about 8:30 P.M. on Friday. "Formal advise" of the failure of the last of the four major power cables supplying the Auckland CBD was received in Wellington by the Minister of Energy, Max Bradford, who gave an account of it in Parliament during a special debate on the outage held Tuesday, February 24 in which he moved that "the House take note of matters associated with the supply of electricity to central Auckland" (Bradford, cited in Hansard, 02/24/98). The Minister of Energy, as well as the Minister of Police, Jack Elder, were both in attendance at the second meeting at Mercury Energy held the following day. Involvement of the central government in the crisis remained limited. This was in part due to the fact that a state of emergency was never invoked.

8. Interestingly, in an interview, Manager Network Customer Services at Mercury Energy, Raymond Cheng, dismissed the likelihood of such a call from Mercury to the Police even taking place.

9. They are actually called Auckland Health are and are part of a hospital organization located in a smaller area of Auckland but they have been contracted

to supply the whole region with public health services. The Medical Officers of Health and the Health Protection Officers working at Auckland Health Care are designated by the Director General of Health at the central level and they have a direct officer-like relation to her or him. These officers can be given directions from Wellington but usually have their own management structure worked out in a wider contract.

10. There generally seems to have been some gaps in the preparedness for entering and working in blacked-out buildings among the Emergency Services. Most of these discrepancies were dealt with during the course of the crisis but others were identified in the post-emergency phase and were incorporated into the revision of the emergency management procedures.

11. For details, see Ministry of Civil Defence, New Zealand (1993) National Civil Defence Plan.

12. Effective utilization of internal expertise within Mercury Energy to secure the external reserve power so badly needed involved reliance on the experience and skills of Paul de Abaitua, the Manager of the CBD Supply Diversity Project at Mercury. De Abaitua was the Mercury employee responsible for obtaining and administering generators desperately needed in the powerless city. He had large-scale private sector logistics operations experience which he brought with him in his response to the crisis. He had previously worked with generator farms in Papua New Guinea, Chile, and Argentina (de Abaitua, interview, 11/16/98). His prior experience with contracting and moving heavy equipment meant that he "knew how to short-cut the system," expediting the secural of aircraft and generators through, among other things, the deposit of money into a London bank account (ibid.) A total of seventy-three generators were brought in by de Abaitua and his team at Mercury to provide businesses and organizations within the CBD with power over the duration of the crisis, most of them shipped to Auckland on Russian heavy-lift Antonov 124 (An-124) aircraft (ibid.)

13. The "U2 plane" which de Abaitua is referring to here is the aircraft which the rock band U2, scheduled for a concert in Sydney when the Auckland power outage occured, used for carrying the band's equipment.

14. The fax stated that "Immediate Strengths" in terms of emergency accomodation included "64 Auckland City Property units currently available; 34 Auckland City Pensioners units currently available" which "[a]t moderate levels of occupation" would provide shelter for approximately 390 persons (Ministry of Civil Defence, Auckland, fax, 1998). Goals of the draft accommodation plan included immediate establishment of an "Accommodation Allocation Centre" (ACC) and in the event of "escalating needs" having ACC halls, ACC community centers, the Sky City Ballroom, other halls, private residences, military accommodation, and local authority assistance respond by providing shelter (ibid.)

15. The lack of precision in this figure is due to the fact that the Auckland power crisis occurred between censuses (Henare, interview, 11/16/98).

16. It is noteworthy that Hugh Rennie is also the Chairman of the foreign-owned operator of the New Zealand Electricity Market (NZEM), the Marketplace Company (M-co.), referred to in the section on electricity reform.

17. The exact Terms of Reference of the Ministerial Inquiry were as follows: "The Government considers it important to carry out a review of the recent failures of Mercury Energy Ltd.'s cables to the Auckland CBD in order to establish what happened, why and lessons for the future. To this end the Government has established a Ministerial Inquiry to publicly inquire into and report to the Minister of Energy before the end of June 1998 on the following questions: 1. What caused the power supply, in particular, the power cables into the central business district of Auckland, to fail? Factors to be considered include: organisational issues, including all aspects of governance structures, accountability arrangements and customer contracts; risk management and contingency planning; technical factors, including network design and quality, operating standards and practices, maintenance standards, and asset management policy and practices, and the weather; and any other relevant factors. 2. In light of the cable failures, to recommend changes to ensure that security risks relating to Mercury Energy's distribution lines are managed efficiently. Any recommendations should take into account the costs and benefits of any changes, including the costs of alternative measures to avoid or ensure against outages. In undertaking the review the Inquiry shall: (a) as soon as possible publish a report prepared by technical experts commissioned by the Inquiry on the technical factors in question 1 above; (b) call for public submissions following publication of the technical report; (c) not determine issues related to liability, breach of law, or compensation; and (d) decide its own procedure" (Ministry of Commerce, New Zealand, 1998:21–22).

18. The Commission of Inquiry stated that they then enlisted the English company, EA Technology, to review the Integral Energy report (Minstry of Commerce, New Zealand, 1998).

19. The remaing submitters regarding technical causes of the cable failures were Brian Leyland, a consultant and engineer who was involved in the original installation of the gas cables; W. R. Braithwaite; Engel Engelbrecht, principal at the firm Connell Wagner; John Collinge, former AEPB Chairman from 1980–1992; and R. B. Arlidge and Associates. In addition to Mercury, those who submitted on organizational issues included the AECT (Mercury's majority shareholder), Auckland City Council, Mark Green, and former Trustee and Mercury Director Peter O'Brien. Operations and maintenance issues were examined along with a submission from John Collinge, as well as an unvalidated report from Delmech Engineering, accepted with reservations by the Inquiry as it was commissioned by Mercury far into the proceedings and thus, had not been reviewed. Risk management and contingency planning issues were evaluated along with submissions from the Auckland Regional Council and John Gardenier (an engineer and risk assessment consultant). John Collinge, Auckland City Council, Robert Western (former AEPB Board member and Chairman of the Engineering Committee from 1983–1990 as well as a founding trustee of the AECT) and Peter Hulme presented submissions to the Inquiry on the issue of Network Distribution Planning (Ministry of Commerce, New Zealand, 1998).

20. There were however, some complications with the nearby residents who had not been specifically informed. As the work with putting up the poles went on day and night, this caused some disturbances for the people living close to the railroad (MoCD, 1998:27).

4

Urban Crisis Management in Auckland: Thematic Analysis

4.1 CRISIS DEVELOPMENT

The four successive cable failures in the complex underground electricity grid supplying the CBD of the city of Auckland began January 22 and thereafter occurred on February 9, February 19, and February 20, 1998. It is not within the scope of this study to perform a full-fledged causal analysis of why this disaster occurred. However, our findings may shed some light on the processes of crisis development.

Despite the element of surprise in the complete failure of all main cables supplying Auckland which has been repeatedly emphasised by Mercury Energy, there were preexisting problems with the electricity system. Some of these concerns were known and documented prior to the outage and came up in inquiries following the crisis. Mercury Energy's own report of July 1997 for example, entitled *CBD Risk Management*, details both Mercury's serious concern about the fragility of the CBD network and the measures the company deemed appropriate to prevent a collapse—measures which unfortunately proved insufficient.

Two of the four main cables which comprised the electricity transmission system on which the Auckland CBD relied were rather old, although they were still within their estimated operating life. These were the first two cables to go in the sequence of failures. Both the No. 2 Penrose-Quay cable which failed first on January 22 and the No. 1 Penrose-Quay cable which failed next on February 9 were 110 kV gas-filled cables (impregnated pressure (IP) cables) installed in the Auckland transmission network in 1959 (Ministry of Commerce, New Zealand, 1998). In addition to their age, it was well-known within the international engineering community since the 1960s that IP gas cables in particular proved problematic—so much so that the manufacturer had in fact, ceased producing them:

> Gas cables have become associated with gas leaks and lengthy repairs. Although IP gas cables were withdrawn from manufacture in the 1960s, those that remain in use can continue to provide reliable service if they are operated within their design limits and if they have been correctly pressurised over their lifetimes (Ministry of Commerce, New Zealand, 1998:53).

Mercury Energy's risk management report of 1997 had declared the security[1] of the CBD system "severely compromised" by the unreliability of these gas cables (Mercury Energy Limited, *CBD Risk Management,* cited in Ministry of Commerce, New Zealand, 1998:44). Mercury's approach to improving the threatened security included replacement of a component cable, installation of new links and an inter-tie between the Victoria and Hobson substations, as well as upgrading other areas of the system (Ministry of Commerce, New Zealand, 1998). The other two cables supplying the CBD, the Roskill-Liverpool 110 kV oil-filled cables, had been installed in 1976 (Ministry of Commerce, New Zealand, 1998). According to calculations of engineers at Mercury Energy, the newer cables, combined with the two older should have provided a network sufficient for meeting the electricity demand of Auckland's city center for some time.[2] Supplementing the system with the addition of two new cables in 1976 however, increased the complexity of the supply grid and thus the number of interactions possible in the system.[3]

The power supply grid that supplied Auckland was tightly coupled, i.e. there was little flexibility in the system or opportunity for doing things independently or differently; there was little room for adaptation to unplanned circumstances (Perrow, 1999). In contrast, an error or problem in a loosely coupled system allows more opportunity for "jury-rigging" or the organization of alternative solutions in a less restrictive operating space and time frame (ibid.)[4] When a system such as the Auckland power grid is tightly coupled, initial failures which occur often "cannot be contained or isolated and the system stopped; failures will cascade until a major part of the system or all of it will fail" (Perrow, 1994:216).[5] The interrelation of system parts in the Auckland power grid resulted in a cascading of failures, beginning with the older gas cables and ending with the collapse of the entire system. As each cable failed, the load on the remaining cables increased—making further failures more likely.

In addition to gas leaks that were recorded very early on with the gas cables, there had been forty-seven nonelectrical faults on the cables within a period of just over thirty-three years (Ministry of Commerce, New Zealand, 1998). The occurrence of faults increased from the 1980s, as did contractor damage (ibid.) The first electrical fault occurred in April of 1995 on the No. 2 Penrose-Quay, the second on the No. 1 Penrose-Quay cable in December of the same year and the last prior to the ultimate system collapse occurred on the No. 2 gas cable again in February of 1997. It is worth noting that no electrical faults had occurred on either of the Roskill-Liverpool 110 kV oil cables prior to the 1998 failures, and a total of nine nonelectrical incidents over the thirteen year period prior to the Auckland outage of 1998 (ibid.)

The significance of the aforementioned factors as well as the obvious expansion of the Auckland metropolitan region was not missed by engineers in the former Auckland Electric Power Board (AEPB). Options for reinforcing the CBD network had come up for discussion there as early as 1988. It was ultimately agreed that a tunnel would be the preferred reinforcement option for the area (Ministry of Commerce, New Zealand, 1998). Somewhat belatedly, the 1993 successor to the network, Mercury Energy, approved the tunnel option in 1995. The three-year project began in 1997 following governmental approval (ibid.) Given the fact that Mercury Energy was however, apprised of the weaknesses and the potential for problems in the supply grid, particularly with respect to the unreliability of the older gas cables, it is remarkable that confidence in the system—at least on an operational level—was high. The ministerial inquiry following the outage found that "As with the gas cables, Mercury Energy maintained that the oil cables could be safely operated up to their full ratings without risk of electrical failure until the failures of early 1998" (Ministry of Commerce, New Zealand, 1998:71). While maximum ratings were not exceeded prior to 1998, these ratings were determined by the transformers rather than the cables. The assumption appears to have been made by Mercury Energy operators that "cable loads, at the transformer maximum capacity, would be well within the cable maximum capacity. This, in hindsight, has shown to be incorrect" (ibid.:68).

Perrow's "Normal Accident Theory" (NAT) extends the narrower interpretation of coupling applied in engineering to include factors external to systems yet capable of exercising decided impacts upon them (Perrow, 1999). The heat sensitivity of the cables in Auckland coupled with elevated temperatures and thermal conductivity as well as demand provides a further example of tight coupling. The New Zealand summer of 1997–1998 had been very warm—it was "the eighth warmest since records began in 1968" and "soil moisture calculations suggest that central Auckland soil was dried out," a factor that increases conductivity (Ministry of Commerce, New Zealand, 1998:62). There was a very sharp increase in electricity demand in January which is reflected on plots of all four of the cables.[6] As noted in earlier sections, Mercury officials had attributed this in part to the increased usage of electrical fans and air conditioning. In sum, several key aspects of the crisis development trajectory in this case coincide with elements of Perrow's Normal Accidents Theory. Further study of the causes of this crisis is necessary however, to develop a more complete and systematic diagnosis.

4.2 PROBLEM FRAMING

The way in which the crisis actors define the problems at hand has important implications for the evolution of the crisis management in a given case (Sylvan and Voss, 1998). Depending on how the problem is framed, different aspects of the crisis are prioritized—often to the neglect of other aspects. If the crisis is seen as a public confidence problem for an actor (for example, a government agency), information management, media relations, and symbolic actions are likely to become key priorities. The framing of the problem also tends to define and determine who is in charge and whose responsibility it is to develop a response (Sundelius, Stern, and Bynander, 1997).

Problem frames push decision makers toward certain solutions to the problem and away from other alternatives. It should also be noted that framing a problem as a "crisis" or "emergency" may change the institutional rules of the game and justify the use of alternative decision making procedures and resources (Edelman, 1988; Moens, 1990).

There are a number of interesting examples of problem framing and frame shifts in the Auckland case. The precrisis framing concerned electrical system security in terms of maintaining the integrity of the infrastructure and reliable access to the power supply. The emphasis in this initial framing was on the improbability of a major system failure. One of the reasons why the management at Mercury initially acted the way they did when the first cable malfunctioned had to do with this initial problem framing and the consensus behind it. There was massive consensus in the industry that the simultaneous failure of four major power cables serving a city like Auckland was virtually impossible (Integral Energy, 1998:13–17). There was, consequently, no contingency plan for what should be done in the event of a quadruple cable failure or any assessment of what the overall consequences of such a failure would be for the consumers and for the system as such. This being the initial mindset of the management of the power company, it is somewhat less surprising that the warning signs that existed (such as repeated smaller power disruptions, and the few critical voices that allegedly resided within the company) were not given adequate attention nor sufficient credit.

At least some representatives of management at Mercury expressed that they had not felt seriously alarmed even when the third cable broke down. This development was seen as quite extraordinary and inconvenient, but some of the management appeared to still be convinced that the fourth cable would remain intact. For example, at the Tuesday, February 17 meeting, two of the cables had collapsed and Mercury Energy informed representatives from the City Council that there may be "some outages" as a consequence and asked them "to be a little bit more economical with [their] use of power" (Manley, interview, 11/17/98). According to one representative at the meeting the crisis was still a matter of a power cut (an ordinary power problem) and there was no indication from the company that this was something major (ibid.) A situation where all cables failed within a week simply could not happen, that is, in the minds of those working at Mercury (Cheng, interview, 11/16/98). Given this mindset, the actions that were taken as countermeasures (i.e. the power saving that Mercury asked CBD clients to adhere to and the repair work that was initiated following the first cable failures) did not seem unreasonable nor quite as insufficient as they might with the benefit of hindsight.[7] A similar problem-framing situation confronted the Carter administration when the Iranian Shah lost his power in 1979. During the political turbulence preceding the fall of the Shah, the only question posed in the White House for a long time, was "How could the Shah be strengthened?" Nobody asked, "What do we do if the Shah falls?" (Moens, 1991:211–37).

In relation to the possibility of the power crisis occurring in the way that it did, some interested parties argued that this was not as "impossible" as the power company was making it out to be, and the Government was being about the risks involved. Many Labour Representatives in Parliament framed the problem as one resulting from the recent electricity reforms, other reforms pending (desired but not yet implemented), and Mercury's monopoly (in relation to industry regulation and accountability). This definition of the problem was aired during the parliamentary debate about the outage held on February 24. Mercury however, asserted its industrial compliance several times in public statements made during the first weekend, with Gilbert arguing that "[t]he issue is not one of standard, but the extraordinary level of near simultaneous fault" (Gilbert, cited in Mercury Energy Limited, 02/21/98:12:00 P.M.).

The power crisis was initially framed as a short-term to medium-term problem, which in turn affected which measures were considered and which were not. Mercury Energy stated in its press release on Saturday, February 21 that the CBD would be "without power for the next seven days." This proved to be an important miscalculation:

> When Wayne Gilbert phoned me on the Friday night, he said the power was going to be off for two or three days. Well, it was nearly eight weeks. So, and admittedly that's perhaps a harsh judgement because this was within an hour of it happening, and they really weren't certain of the extent of the problem even then, but they did not, in the early days understand the magnitude of their problem. As I say, if I was critical, it was that they were unduly optimistic for much too long before they came clean and said look, we are not going to be able to fix this—the only solution now is the line in from Penrose. (May, interview, 11/18/98).

The initial short time frame seemed more and more unrealistic as time went on and the cable tests continued to fail, something that contributed to a new perspective of the problem as something more complex and long-term.

In the acute phase at the outset, a key framing issue was whether or not there would be a declaration of emergency. As we have seen, there were strong advocates on each side. Some thought that use of the emergency legislation would provide structure to the crisis response. The Mayor, however, was strongly opposed to a declaration and his view was supported by the emergency services' assessment of the situation as potentially dangerous but under control. In Auckland, there was no official declaration of emergency, but the situation at hand clearly required complex, inter-organizational cooperation and vigorous action. Many decision makers experienced the crisis as taking place within a grey zone of legal authority. Given this uncertainty (and the coordination difficulties associated with it), various ad hoc coordination mechanisms gradually emerged to facilitate a more integrated crisis response and crisis communication approach (cf. 't Hart et al., 1993).

It became clear towards the end of the first week that the rescue services, although they were quite busy, had the situation under control. Some people were trapped in buildings. Others suffered heart attacks from the heat and from walking up long stairways. However, nobody was seriously injured and nobody died as a direct consequence of the loss of power. The focus then shifted to the losses that the businesses in the area were suffering and would be suffering as long as the power was out. The Minister of Finance, Hon. Bill Birch, mentioned the greater economic impact of the crisis as early as Monday, February 28, stating that "[a]t this stage, as with the Asian crisis, the economic impact of the power crisis is unclear. A great deal depends on how fast Mercury Energy can get its systems up and running" (Hansard, 02/24/98). This framing—with a primarily economic focus—was widened as time went by, to include the losses the CBD as a metropolitan area was suffering and eventually, the economic impact Auckland as a whole was feeling as a consequence of the CBD being virtually shut down. The international aspect (with the Australian travel sector discouraging travel to Auckland in general) highlights the type of wider economic problems that greater Auckland came to experience as a consequence of the power outage in the CBD.

The final shift in framing came when the power outage in the CBD was no longer seen as primarily a public safety problem but as a local economic crisis. There seemed to be a strong consensus among the people involved that this shift in problem framing took place at the end of the first week of the outage (Manley, interview, 11/17/98). Virtually none of the worst case scenarios materialised that the police and rescue services had envisioned and prepared for. Yet other problems—such as the economic consequences—came to the fore as the crisis wore on.

4.3 LEADERSHIP AND GROUP DYNAMICS

Early on in a crisis when much is still up in the air, there is likely to be considerable uncertainty about the nature and scope of the crisis. Jurisdictional and resource issues are likely to be clouded with uncertainty as well. These issues are tightly coupled with the initial problem framing and "Who owns the problem?" tends to be an open question. In this initial void there are substantial opportunities for an actor either to appropriate the problem, thereby getting to shape the management of the crisis, or to "pass the buck" and avoid taking responsibility. Such an avoidance strategy could involve for instance, framing of the problems in such a way that would make another organisation the natural lead agency. Leadership plays into this issue in a number of ways. An actor that shows strong leadership or that is able to take initiative early on in a crisis, may be seen as a natural leader for the overall management of the crisis and may thereby be given a great deal of responsibility in the crisis. When a crisis develops well and is generally considered a success in the eyes of the public, taking initiative, showing resolve and being in charge is the dream position of any politician, company or agency. It can provide political and commercial opportunity with lasting future benefits. In cases of policy fiascos and disasters that go from bad to worse, being in the lead position can mean heavy loss and can cause great personal stress as it may mean the end of a careers and even the "death" of discredited companies or agencies (Bovens and 't Hart, 1996). Accountability issues mirrored in the press often depend not only on whether the actor who managed the crisis did a good job but also on whether the actors the public think should have had a key role actually assumed responsibility for the management and showed good leadership.

Leadership issues are of interest, not only in regard to who appropriates the crisis or alternatively, who gets stuck with it, but also in regard to what happens within the groups that assemble to manage a crisis, especially when people in a particular group are not used to working together. Scholars on group decision making in crisis situations sometimes talk about "newgroup syndrome" (Stern, 1997; 1999). This concept refers to a pattern in the formation stage of a decision making or advisory group, which promotes conformity within the group:

> In ad hoc or newly institutionalized groups . . . a common group subculture and well-developed procedural norms tend to be lacking. This vacuum creates uncertainty among the members, which works to the advantage of group leaders and other assertive individuals and contributes to the emergence of tendencies toward conformity in the group as a whole (Stern and Sundelius, 1997:127).

Leadership also concerns the decision group and its contact with the outside world—with other decision groups, the press, the public, or victims of the crisis. In this contact, certain people sometimes get a function where they become so-called gatekeepers (Hoyt, 1997) who channel the flow of information from one group to another.

The power crisis in Auckland became something of a game of Old Maid almost from the second it occurred. The Old Maid is the playing card that everybody tries to avoid getting stuck with by passing it on to another player. This was very much how responsibility for the crisis was treated early on in the crisis. Mercury Energy called the police and wanted a state of emergency declared, which would have passed the responsibility for the crisis largely into the hands of the central government.

After some initial passing of the buck Mercury ended up with the Old Maid in its lap for a number of reasons. Assuming responsibility for what had gone wrong was not initially seen as having anything to do with leading the management of the crisis but Mercury Energy eventually realized that it had to do both. The leading role that the company took came a lot earlier than the acceptance of the responsibility for what had caused the power crisis. The management at Mercury Energy took charge of the situation very early on by calling the meeting on the Friday evening and the Saturday morning. They were the ones that had the most information about what was going on and they called the other responsible decision makers to inform them of the situation. It has also been noted by several actors involved in this case that the early initiatives and the resolve showed by Mercury at the start of the crisis gave the company a privileged position in the management of the crisis that it may not have had otherwise (Heavey, interview, 11/18/98; Hutchinson, Small, McKeen, and Wood, interview, 11/18/98).

Mercury Energy stepped into the central role as the primary agency (with the available physical space and resources) through which management issues were shaped, discussed, and coordinated. There may have been several factors in conjunction that resulted in the crisis being lead by Mercury. For one, Mercury displayed strong leadership early on in the crisis and secondly, as there was no declaration of emergency, there was no obvious agency or agency representative to take over this leadership role. Mercury set up the operations centre and pulled together the joint management group that met at Mercury's office twice a day throughout the duration the crisis.

In the joint management group assembled during the Auckland outage, various characteristics of "newgroup syndrome" were visible. The fact that it was a new decision-making constellation was obvious. It was an ad hoc group put together by Mercury who invited all the actors who had a stake in the crisis. Name tags and passes were issued and signs had to be put up throughout the Mercury offices so that everybody would know where they were supposed to go. Many reported an initial lack of common rules for how the meetings should be set up, how many representatives each organization could bring and a lack of common meeting etiquette. At the first meeting the confusion concerning what was supposed to take place, who was in charge and what the rules in the new group were was so great that the whole group (as is quite common in such situations) fell completely silent. After having waited in silence a while one of the more assertive group representatives, stood up, banged his fist on the table and asked loudly, "What is going on here? Who's running this meeting?" (Meeting participant, interview).

During the first couple of meetings Mercury Energy took charge and set up the meeting format (i.e., standardized rules on who could speak, when, and for how long, etc.) without much protest or overt objections from the others present. The circumstances surrounding the formation of the joint management group at Mercury Energy seem to support the presence of so called "newgroup syndrome." "Newgroup syndrome," aside from privileging assertive parties in a group, leads to a less critical scrutiny, at least overtly, of the decision-making process (Stern, 1997:153–91). What is also noteworthy from this perspective is the strong consensus in the group and support for Mercury's, the City Council's and emergency services' action plans for the crisis. These can be said to have been the three most assertive actors in the management group and they enjoyed considerable influence in the decisions made within the group. Whether this had a negative effect and in some way led to poor or less creative decisions in the Auckland crisis is impossible to say. There are no indications that the decision-making process deteriorated as a consequence of the strong consensus which emerged, there is only the theoretical possibility associated with the type of group behaviour observed in other cases, such the Bay of Pigs invasion (Janis, 1982; George, 1980) and the U.S. rescue attempt in Tehran (Smith, 1985).

Mercury Energy was also the absolute focus of attention from a practical point of view, as the instigator as well as the "fixer" of the problem at hand. Once the company ended up in the driver's seat, having a virtual monopoly of expertise in the area of electricity strengthened its management position.[8] The quick and exceptionally functional setup of the operations center at Mercury also played a part in the leadership role that the company acquired.[9] Cheng's effort to adjust the Mercury office building into an operations centre where all of the actors could run their operations very smoothly at a strategic level strengthened Mercury's role as the central coordinating body—a role to which there was no feasible alternative under the circumstances.

The leadership at Mercury had two faces, one at the Mercury meetings and one directed outward towards the media. Inside Mercury the operative frontline manager, Raymond Cheng, and Chief Executive Wayne Gilbert played key roles. Externally, Mercury's Board Chairman Wayne Donnelly, the General Manager Network, Richard Gibbons, and Chief Executive Wayne Gilbert represented Mercury Energy. A few weeks into the crisis, the leadership at Mercury shifted a bit, which may have weakened the company's position a little. This probably put additional stress on the remaining members of the management team. After a skirmish Wayne Donnelly resigned from his post as Chairman of the Board, due to what could be considered a conflict of interest between his role as a director on the City Council (as well as acting Chief Executive for the City Council for the first couple of days) and his role at Mercury Energy. Mercury thereby lost a person with considerable skills in media management and public confidence building. Dr. Patrick Strange, the company's chief engineer, then stepped in as Chairman of the Board and stood side-by-side Wayne Gilbert in the increasingly burning spotlight of media scrutiny.

The fact that Mercury Energy ended up doing much of the response coordination during the crisis gave the company certain benefits, but Mercury management also became increasingly aware of the downside to this high-profile position. As the crisis dragged on, Mercury was increasingly seen as the actor responsible for what was experienced as bad management of the crisis as well as being seen as having caused the initial problem.

Overall, there was felt a lack of political and societal leadership in the crisis. The mayor would have been one of the natural leaders that many people would have liked to see take a leading role in the management of the crisis. His absence at the leadership front was noted. Those who mentioned the absence of a civic leadership figure have stated that perhaps a senior official from the City Council would have been an appropriate possible alternative to the Mayor in this role (Heavey in interview, 11/18/98).

The Mayor, Les Mills, decided early on to give Mercury Energy his full support during the crisis. He was allegedly both annoyed and worried about the fact that the city's largest power company had let the situation become what it had. Because the Mayor and his advisory group felt that Mercury had caused the crisis and it was up to Mercury to fix it. He ignored the public political pressure that was put on him as a political leader. Mills and his advisers ignored questions and accusations confronting the mayor (such as: "What he was going to do?;" "Why did he let the situation come to this," and "He should not have let this happen"). The Mayor was nevertheless determined to stand by Mercury throughout the crisis. He stated that the company was doing everything it could to fix the problem. This was a decision he made on his own initiative and it turned out to be a politically costly one. Lending his credibility to Mercury and becoming publicly associated with Mercury was what probably cost him the 1998 mayoral election later that fall (Manley, interview, 11/17/98).

The agencies that almost exclusively benefited from the strategic leadership vacuum and their own operational assertiveness were the Police, the Fire Service, and the Public Health authorities. As earlier mentioned the crisis on the whole turned out to be a great PR success for the operational services.

4.4 ORGANIZATIONAL COOPERATION AND CONFLICT

Crises present challenges of political consensus building and intra-organizational bargaining. National crisis decision making in the United States has often entailed an isolation of the political leadership in which the participants are restricted to a fairly narrow organisational and political base. This may be different in more consensual political systems (see Bovens, 't Hart, and Peters, 2001).

In Sweden, for example, during some of the most serious foreign policy crises, the political leadership has included all of the political parties in Parliament and there has been a conscious striving towards consensus building within this group. One advantage of this approach is that the external pressure seems to decrease in these situations in comparison to when one or a few decision makers make a decision. Critics can also be included in the decision-making group and thus be politically disarmed by being forced to share responsibility. With the strong "anchoring" of decisions among the political parties, a broadly supported policy can be presented to the public. Taken to its extreme, this consensual style of crisis management can however, also result in nonsubstantive decisions and a lack of, or even an evasion of, critical analysis (Sundelius, Stern, and Bynander, 1997:150–51).

The Auckland case displayed a combination of the two prototypical decision-making styles. There was a strong striving towards consensus around which measures should be taken and a collaborative approach to defining and solving problems. The interagency meeting arrangements at Mercury Energy testify to this ambition. The Minister of Police, Jack Elder, described what he saw at the first Mercury meeting in the following manner:

> One of the things that has impressed me most about this problem [the power outage] as it arose was the cooperation of people in Auckland. That meeting on Saturday morning, which Mr. Bradford [Minister of Energy] and I attended, was attended by a whole load of other people who represented different interests in Auckland. . . . Absolute cooperation was shown by everyone. There was not a whole lot of carping and whinging—there was absolute cooperation. They set about doing something about it (Elder cited in Hansard, 02/24/98).

The meetings at Mercury also illustrated that consensus (or at least a wish to avoid exclusion of interested parties) was important since all actors were invited to partake in the meeting even if the number of representatives from each agency had to be decreased quite quickly. The way the meetings were set up and run at Mercury seems to have been a solid ground for legitimizing decisions and forming a unified policy line to be presented to the public. The expert monopoly that Mercury Energy possessed most likely facilitated the building of a unified problem frame and solution package. Despite these factors supporting the consensus building there is evidence that this did not work perfectly. One of the first displays of dissent seemed (contrary to Mercury's official account) to have come when employees at Mercury apparently were making contradictory statements about issues pertaining to the outage. Mercury quickly firmed up its media policy and informed their staff that all questions about the outage should be directed to the media liaison officers working at Mercury.[10]

Striving for consensus at the Mercury group meetings moderated but did not fully alleviate interorganizational tensions during the crisis. Emergency Services, as noted earlier, worked very well together. This was partly thanks to the repeated exercises the services had performed together. For this reason, the officials involved knew each other quite well. Moreover, these three organizations regularly work under very similar constraints and with similar priorities. These three groups did not find the Mercury group meetings very useful in terms of helping them solve their operative tasks. Eventually the Emergency Services set up a separate meeting group where they met every morning before going to the Mercury Energy meeting. Emergency Services did not find a need to discuss the operative issues and decisions they were working on in the Mercury forum. For better or for worse, the larger, more strategic management was excluded from the decision-making process between the three groups concerning the most immediate response. They were merely kept informed about how things were developing in the crisis response.

There may have been other additional reasons why the Emergency Services were not too keen on the Mercury meetings. Mercury Energy was running half of the meeting and the City Council was running the other half. The Police however, was leery of Mercury from early on. They had received information at the start of the crisis that seemed to contradict the information Mercury Energy was providing them with. This information discrepancy continued over the course of the crisis, which undermined the purpose of the meetings in the eyes of the police.[11] The meetings that Emergency Services held among themselves were therefore much more appreciated by the Police who then mainly utilized this forum rather than the meetings at Mercury Energy.

Two other actors were also initially having problems cooperating. The Auckland Public Health authorities and the City Council did not get off to a really good start in the crisis. The Environmental Health Officers on the City Council were the natural cooperation link between the health authorities and the City Council. Environmental Health Officers have the same training as the Health Protection Officers working at the department of Public Health. Environmental Health Officers and Health Protection Officers basically do the same type of job only in different organizations. One difference however, lies in that Health Protection Officers can work across the entire country in an emergency situation because they are warranted by the Director General of Health. The health authorities and the City Council needed to work together for two reasons: (1) Public Health needed backup personnel who had training in the field and (2) the City Council needed the support of Public Health's authority. If premises needed to be closed or if improperly-handled food needed to be confiscated, only the health authorities had the mandate to do this. Two logistical issues made it difficult for the Health Authorities and parts of the City Council to cooperate, at least initially. The City Council building had been cut off from the power supply and related to this problem which had made working effectively on the premises impossible, the Council had partly relocated outside the affected area and into a new building in the CBD. This made it very difficult to get hold of the necessary people within the Council.

The fact that the decision-making structure of the City Council had transformed into an emergency management structure was confusing for everyone who had to work with the Council. The general rule was that everyone should operate as usual, since there hade been no declaration of emergency. However, the Council did change its structure such that it operated in emergency mode. This caused some confusion and several problems, particularly between the Public Health authorities and City Council (Hope, interview, 11/18/98).

Other parties who also had problems cooperating and coordinating throughout the crisis were the mayor and the City Council. First of all, the Council had a hard time locating the mayor at various times during the crisis. One example of this was when the crisis first hit and the mayor could not be reached to be informed of its occurrence. He was first notified of the outage when the press confronted him. The mayor's office was then moved to Mercury Energy where it became more and more obvious over time that his staying there was not good for his public relations. Mayor Mills then, towards the end of the crisis, spent quite a bit of time in his car, maintaining a sort of mobile office. He had his cellular phone with him, which proved to be of little use for those who wanted to reach him quickly, as it was engaged much of the time.

It was difficult for the Council to, on the one hand, keep Mayor Mills informed of what it was trying to do. At the same time, the Council had difficulties keeping itself informed of what Mills was publicly saying that the Council was doing (Lindsay, interview, 11/17/98). The mayor was known as a person that was not afraid to walk up to the press and make statements or comments about issues based more on his personal beliefs and opinions than any general Council policy line. He had a strong sense of community thought about a number of issues, including quite controversial ones.[12] Council Representatives would catch the Mayor on TV or in the newspapers making statements about policies and responses that had not yet been decided on by the Council. The Council then faced a sort of fait accompli on certain issues where they thought it was best just to go along with the Mayor's voiced policy line despite the fact that they may actually have preferred a different approach.

4.5 CRISIS COMMUNICATION AND MASS MEDIA

The media play an increasingly important role in contemporary crisis situations. The media contribute to the creation, form, and termination of a crisis. Since so much of a crisis has to do with perception, what creates media attention to a great extent determines what is considered a crisis. Not only do media contribute to the shaping of crises, they also have the ability to get to the news before the decision makers do, as happened in this case. Effective crisis communication has become an essential part of crisis management. Actors in crises today are often acutely aware of the fine balancing act of keeping the media at a workable distance and at the same time not appearing to avoid the media or appearing passive in the management of the crisis. Previously, decision makers could keep the media at a distance by having closed negotiations or by referring to a threat as a national security issue or fears of public panic.

Another way in which decision makers sometimes get in trouble with the media is by falling into so called credibility traps. It is not altogether uncommon that ill-advised early statements lead to a lack of, or a deterioration of, credibility at a later stage of the crisis. Overly optimistic statements early on in a crisis can pose a clear risk for decision makers. One of the classic examples of this was Neville Chamberlain's optimistic statement of "Peace in our time," following his return from negotiations with Hitler in Munich. Chamberlain's political credibility soon expired and the much less optimistic Winston Churchill replaced him (Sundelius, Stern, and Bynander, 1997:203–4).

A third aspect that is interesting in regards to the effects of mass media on a crisis is the portrayal of a crisis in international media. Hordes of international journalists typically descend upon the scene of disasters. Problems associated with "disaster tourism," (Rosenthal and 't Hart, 1998:200), where journalists and curious people crowd the operative site and in certain cases hinder emergency relief agencies from performing their tasks, is being amplified with increased communication and mobility. Another part of the "disaster tourism" which is also increasingly taking up the valuable time and energy of decision makers in crises are VIP visits, where if one national politician or official person visits the disaster area lots of others feel they have to go as well. These VIP visits may very well be helpful in providing moral support for the community hit and in underscoring official promises of support but they also require a lot of attention on the part of the crisis managers. In part this is tied to political opportunism related to crises, where showing empathy on site can be a serious PR boost and failing to do so at the right time may mean the end of a political career. One example of this can be found in the 1975 stadium disaster, in Brussels, in which the Belgian Minister of the Interior chose not to go to the stadium and get directly involved in the crisis and instead entrusted a subordinate to act for the ministry in the crisis. Despite the Minister's rationale to not get in the way of the emergency services and disturb the on-sight crisis group's management of the disaster, he was severely criticised for his lack of leadership and his failing to show empathy in the post-crisis phase and the question of whether he would need to resign became a heated issue in parliament ('t Hart and Pijnenburg, 1989).

Looking at the Auckland case, Mercury Energy created a credibility trap for itself. There was a buildup of expectations each time Mercury stated that a cable was ready for testing. When power subsequently was not restored (as a result of failed tests), Mercury's credibility suffered. An approach that was supposed to convey honesty and pragmatism backfired. It just made the company look not in charge (at best) and incompetent or even deceitful (at worst). Mercury was very well aware that the possibility of the power being restored the next day depended solely on a successful test result, but somehow the public did not grasp this. The progress reports (that were continuously revised together with tentative time schedules for power restoration) created a credibility trap that resulted in serious PR damage for Mercury Energy.

Mercury and the Auckland Police had quite different crisis communication approaches. In part this was by choice, but it also sprang from the different positions the two parties occupied during the crisis. Mercury can be described as having a more reactive media management style in many respects. The company tried very hard to have an open media policy, holding press conferences and releasing news updates by fax to all of the major newspapers and news agents. Media management, however, does not only entail giving the media one's information. It is also about image management. Mercury did not have a very proactive grip on the negative images being portrayed through the media. They discovered too late that there had been a problem with their communication about the progress updates. Mercury's remedial actions came, simply, too late. The company was targeted as the main culprit for the power outage. And so a visious circle emerged: the very fact that the credibility trap made the company look incompetent only fueled the public's hostile attitude towards Mercury.

The police, in contrast, maintained a proactive approach to crisis communication. Being hardened by regular negative media attention, the police force was determined to dampen any negative feedback in the press by clearly showing from the start all of the measures it was taking to ensure the safety of citizens and property in the CBD. Possibly the police leadership saw the positive opportunities the crisis offered for presenting itself in the media. This was perhaps why they invited the media to their first strategic meeting at the police headquarters. This initial invitation to the press turned out to be the first part of their successful media management approach.

The tale that the mayor's PR advisor, John May, tells about how the Mayor found out about the Auckland power outage shows how the media often get the news first. May was at home on the evening of Friday, February 20, 1998 when he received a call from Mercury Energy's Chief Executive Wayne Gilbert. Gilbert told May that the city was experiencing a major power failure and that the power would not be restored for a few days. May thanked Gilbert for letting him know. John May found this a bit unusual. An important tradesman in the city would not normally call a PR manager at home but would instead talk to the Chief Executive at City Council in this kind of an event. May then deduced that Gilbert may not have been able to get hold of the Mayor. He then proceeded to call his Chief Executive, only to find out that he could not be reached because he was in Australia.

The next question then arose, who was acting as the Chief Executive's stand in? May did not know. He decided to call one of the directors to find out who the acting Chief Executive was. He contacted Bob Wakelin, the Director of Enterprise Services at the council and briefed him on the situation. It was then decided that May would call the Mayor and inform him too. The only problem was that May did not have the Mayor's unlisted telephone number at home. He called the customer call center at the City Council but they did not have the Mayor's number either. May then decided to call the Mayor's Chief Assistant to ask her for the Mayor's home phone number, but he did not have her phone number at home either. May then called the customer call center again, only to discover that they did not have the Mayor's Chief Assistant's number either. Around this time, May recalled that the Mayor would not be home in any event. May recalled that the Mayor—who owned a well-known gym—was coaching an elite discus thrower who was scheduled to compete that night at the North Shore stadium. May then tried to get a hold of the Mayor's cellular phone number, which neither May nor the customer call center had.

May had tried calling the stadium for about half an hour but the competition had already started and nobody was picking up the phone. At this point May was getting frustrated and worried but he came up with an idea. He remembered that Television New Zealand was broadcasting the major events of the competition, so he called them up. He said, "This is not an emergency, and don't go out of your way, but if you are talking to your staff over at North Shore stadium, would you ask them to get Les Mills [the Mayor] to give me a ring at home?" He then turned on the 9:30 P.M. news and saw the Mayor in front of the camera and a reporter sticking lights and a microphone in front of him and asking, "What are you going to do about the crisis?" (Hyde, 1998:31) The Mayor's response was virtually "What crisis?" This was of course what May had been trying desperately to avoid (May, interview, 11/18/98). When a Soviet submarine ran aground in the Swedish archipelago in 1981 there was a similar unfolding of events. The Swedish military headquarters first got word of the submarine when a journalist called the Navy Press Officer and asked what the Navy's comment to the submarine was. The Press Officer then had to answer, "What submarine?" The journalist replied, "The one stranded in the Karlskrona archipelago." Then the Press Officer, clearly irritated, exclaimed, "Are you joking with me?"(Hellberg and Jörle, 1984:80)

In the Auckland case there was a serious message escalation in some of the international media coverage. The power outage in the CBD happened at the height of the tourist season. Auckland has a large tourism industry and is a Mecca for sailboats and yachts from Australia and the Pacific region. It is not surprising, therefore, that the overseas exposure to the power outage became of special interest to the central Government with regards to protecting Auckland's, and later more importantly, New Zealand's, image. This latter issue is tightly tied to the fact that the crisis shifted from a civil emergency to a predominantly economic crisis.

Hence during the crisis, there were signals being picked up by actors, from outside the country, that Auckland was not a good place to go to. The following description of the situation in Auckland was found in an Australian newspaper:

> The fire service warned of a towering inferno, fearful that back-up alarm systems could fail and of death-trap car park buildings, full of toxic fumes because of the lack of ventilation. Health authorities warned of rampant food poisoning if people were not careful with perishable food. (Dunbar cited in MoCD, 1998:20).

In part, this was a consequence of the Mayor's policy of telling people to stay out of the CBD, although other actors also made statements that seemed to support a picture of chaos and mayhem. Early on in the crisis, members of the New Zealand Parliament and even the Deputy Prime Minister went on the official records with questions and statements underlining how the event was unfathomable, that "the residents of a city [Auckland] are being held to ransom" (Winston cited in Hansard 02/24/98), that the situation "threatens lives and livelihoods of thousands" (Winston cited in Hansard, 02/24/98) that one of the world's most modern cities had been "reduced to Third World status." (Winston cited in Hansard, 02/24/98). A member of Parliament from Auckland commented that domestic newspapers were adding to this picture by showing of downtown Auckland closed down and a bad place to be (Tizard cited in Hansard, 02/24/98).

For a long time, both Australia and New Zealand have been each other's greatest source of tourists. During the power outage the Australian Department of Foreign Affairs and Trade in Canterbury published a traveler's advisory, telling Australians to be cautious if they had business or other calls to make in the central part of Auckland. The Australian office had gotten their information from the Australian Council General stationed in Auckland. His office was in one of the blacked-out buildings in the CBD and a situation report was passed on from him. The travel advisory was issued much to the New Zealand's Ministry of Foreign Affairs' surprise. It is not common practice for the ministries of either Australia or New Zealand to go out and issue travel advisories. New Zealand had only issued one once and that was on Iraq when the tension between the United States and Iraq was very high (Heanan, interview, 11/20/98). The text that went out was rather modest, telling people to be careful, but the message that got out was that all of New Zealand was blacked out and that people should not go there (Moore, interview,11/20/98). Because of the good relations between New Zealand and Australia, the unusual action was seen as almost an insult.

New Zealand's Department of Foreign Affairs felt that the advisory was quite uncalled for, since only approximately three square kilometers had blacked out and the situation was not directly life threatening. Most of the people traveling from Australia to New Zealand go to many other places than Auckland's CBD district. The Australian division at the New Zealand Ministry of Foreign Affairs began to make sure that until the crisis was over the Australian government would have plenty of information so that they could make an assessment of the impact of the power outage on Australians. The Australian division called the person in charge of coordinating the event at the Prime Minister's Office and he asked the National Civil Defence Officers in Auckland for daily situation reports. These reports were then handed to the Australian division at the Department of Foreign Affairs who faxed them to the High Commission Officer, who in turn gave the information to the Australian government. After some time the travel advisory was revised and the assessment was restricted only to the area affected. Still, a lot of damage had already occurred.[13]

Foreign media reports together with the travel advisory (alongside the actual power outage, of course) seemed to have had a serious impact on tourist revenues in the Auckland region. The fact that the CBD took tremendous losses from lack of tourism during this period is quite clear. Greater Auckland also experienced a serious drop in tourism during 1998. Whether the advisory actually changed a lot of Australians' travel plans or not cannot be confirmed but it was certainly seen as quite unfortunate and unjustified by New Zealand's government and Ministry of Foreign Affairs (Heanan, interview, 11/20/98).

Auckland also suffered from some disaster tourism, mainly in the shape of VIP visits by politicians, representatives of foreign rescue services and a number of researchers interested in disaster management. Several decision makers who took part in the management of the power outage mentioned that the visits required a fair amount of attention in terms of preparations, accommodations and visits for the various offices involved. They also stated, that although this was time consuming and somewhat tiring at a time when time and energy were scarce, they considered the visits important for the sake of generating and spreading knowledge about crisis management.

4.6 INTELLIGENCE GATHERING AND INFORMATION PROCESSING

Mercury Energy, as the electricity provider and proprietor of the failed network had knowledge regarding developments in the power crisis which many other actors in the crisis were not privy to or lacked expertise for understanding. Thus, Mercury's position in the crisis, while not enviable, gave them certain information advantages. Moreover, Mercury Energy was a private corporate actor which allowed a different approach to communicating information than if for example, they had been a public authority with direct responsibility to an electorate. A 1995 amendment to the National Civil Defence Plan constituting Part Seven and entitled "Public Information" details provisions and describes procedures to be invoked during civil defense emergencies in New Zealand. As an emergency was not declared during the Auckland power outage however, these procedures were not instated and information was released and circulated in a more ad hoc fashion by Mercury and local officials.

While problems in Auckland's electricity grid were evident to a number of actors within the CBD prior to the collapse of the third cable as small and short-lived power failures had occurred since January 20, 1998, Mercury Energy elected not to come forward with public statements regarding the network's fragility until the collapse of the second cable occurred on February 9. At this time, Mercury Energy issued the first of many press releases during the crisis. Mercury's contractual agreements with customers did not guarantee service (Ministry of Commerce, New Zealand, 1998) and thus, their responsibility and accountability to clients appears to have been based largely out of their economic interest, i.e., it was in the company's interest to respond to the public as the relation between customer satisfaction and customer retention is strong. The provision of such key infrastructural services by private actors raises preparedness issues necessitating cooperation with local authorities through for example, programs such as Lifelines.

Mercury maintained an upper hand in information dissemination and the organization of the crisis management response, something which the City

Council and Emergency Services sanctioned. This was clearly apparent in the joint press releases that were issued from Mercury's offices. Nevertheless, the civil defense and Emergency Services personnel developed other ways of maintaining control over their respective interests. The officials involved undoubtedly saw the sense in Raymond Cheng's argument for having the crisis management group stationed on Mercury premises: "The latest information in a power crisis comes from the controller. How do I distribute it effectively to other people unless those other people work here with us?" (Cheng, interview, 11/16/98). Still, despite sitting at Mercury and coordinating their activities from Mercury offices, most officials had no intention of relinquishing control over their areas of responsibility. City Council was perhaps the least proactive and assertive in this respect. The Karangahape (K'Road) Business Association was one of the organizations which "looked for stronger civic leadership. They reasoned that the Mayor should have led, demanding answers to his questions and having the information go through his office" (Director, cited in MoCD, 1998:10). Other important actors in the crisis however, such as the Ministry of Civil Defence and the Auckland Police Department were particularly adept at maintaining their own control over information during the crisis.

The Ministry of Civil Defence managed information during the crisis through the production and distribution of situation reports (Sitreps) in Auckland, which were catalogued and addressed to the Minister of Civil Defence in Wellington. These Sitreps documented proceedings of meetings at Mercury Energy and activities associated with the crisis management response. Copies of press releases and other documents released by Mercury Energy or other agencies, were attached to the Sitreps in order to keep the capital informed of local goings-on. All Sitreps issued during the crisis were approved and signed by the Deputy Commissioner of Civil Defence at the time, Frank Arnott. In addition to participating in joint meetings at Mercury Energy beginning Saturday morning following the fourth cable failure, Civil Defence in Auckland also engaged immediately in planning with the Auckland Regional Council (ARC), establishing prioritized recipients of the limited power available, engaging in "worst-case scenario" planning, and ensuring that all civil defence staff were reachable by telephone (MoCD, 1998).

The Auckland Police Department responded to information management of the crisis by "opening a Special Operations room (which they shut down on 2 March) and brought in additional resources. They coordinated their work with the Fire and Ambulance Services" (ibid.) Had a state of emergency been declared during the power crisis, the Police would have led the emergency services assemblage, as prescribed by a framework document entitled the

"Emergency Services Coordinating Committee" (ibid.) As this was not the case, "a committee was set up through Mercury and a secondary set of meetings took place among the emergency services" (ibid.) The Police maintained contact with Mercury Energy through a staff member who provided them with inside information. The daily meetings that the Police Department held with the Fire and Ambulance Services outside of the broadly-attended Mercury meetings were highly valued. There was a degree of distrust of Mercury, particularly among the Police, which made the Emergency Services keen to maintain their own coordination platform. The Fire Service also held "an internal brain-storming session and received reports back from fire crews that visited the central area" (ibid.:13).

4.7 STRESS MANAGEMENT

There are a number of ways in which stress affects decision makers and emergency response teams throughout a crisis. All stress is not necessarily negative and there are a number of factors that function as buffers or mediators in stressful situations that either enhance or decrease an individual's capacity to function (Flin, 1996:98). The relation between experienced stress and performance of various tasks is often described as an inverted U-curve (George, 1986:534), with low to moderate stress potentially improving performance and protracted high stress degrading performance (Holsti, 1989:28). In stressed situations people experience, in varying degrees, the psychological and physiological reaction called "flight/fight-response."[14] The physiological part of this response to stress can sometimes, combined with a confidence that the situation can be solved or managed properly, lead some people under moderate levels of stress to feel a certain "high," or to feel sharper and more effective than ever before (Flin, 1996:98). Even in crisis situations people may display this type of extraordinary capacity. (Holsti, 1989:28). These types of situations can be described as forms of positive stress; stress that increases output in different ways. Negative stress on the other hand, impedes the functioning capacity of a person and is experienced when demands are felt to exceed resources. The experience of negative stress interacts with mental, physical, and emotional factors, which bear different weight for each individual in each situation (Flin, 1996:98;100). There is a substantial body of research that suggests that "intensive and protracted crises may erode rather than enhance the ability to perform these [cognitive, information-processing and management] tasks" (Holsti, 1989:26) and that this change in effective management is particularly linked to the effects of stress (Holsti, 1989:26) The qualitative difference in the required performance by people in strategic leadership positions make these people more vulnerable to negative stress, especially intense and protracted stress. Crises that entail complexities and uncertainty, which is most often the case, and that also require responses that are measured qualitatively, not quantitatively, often produce this type of negative stress for the crisis manager (Holsti, 1989:28).

In the literature on stress another useful distinction is sometimes made, that between chronic stress and acute stress. Chronic stress, also called occupational stress, is linked the workplace conditions and the individual's reactions to these over a longer period of time. Acute stress, also called emergency stress or critical incident stress, is more closely linked to crisis situations (Flin, 1996:98). In its worst form it faces the individual quickly, either in the form of a life threatening situation or perhaps a traumatic event and may cause the "flight/fight-response."

In a crisis situation such as the Auckland power outage it may be helpful to make an initial distinction between actors for whom stress is a routine factor in their work and those actors that do not normally experience the kind of stress they faced in this case. Emergency services personnel typically experience stressful work environments. This often entails negative stress, but firefighters and ambulance staff are well-trained and have ample experience in dealing with the kind of stress their work entails. Collegial support, debriefings, and training exercises are firmly embedded in the organizational culture. A number of factors have been identified as being especially stressful for emergency and rescue personnel in crisis situations (Flin, 1996:103). These include threat to core values such as life and health; threat to one's own life; sense of responsibility and fear of failure; rapid changes needing new assessments; insecurity in assessment and solutions; lack of information or information overload; and group tensions (Flin, 1996:103).

In the Auckland case these factors were largely absent, which may help to explain why the emergency services experienced few difficulties in performing their key tasks, and had such an overall good feeling about the crisis. There was a potential threat to life and health, both for the CBD residents and emergency services personnel, but this was brought under full control during the middle of the first week. There were initial fears that the Fire Service switchboard would break down, that the law and order situation would become unbearable, and that water supplies would not be kept fresh and uncontaminated. These uncertainties decreased quite quickly however, and the threat of failure was virtually eliminated as the work of the emergency services increasingly took on the characteristics of rather intense but routine operations. Overall responsibility for crisis management had come to rest with Mercury Energy, which helped decrease any concern linked to immediate responsibility. The information problems that occurred were remedied by adding extra meetings among the three emergency services.

There were also actors that are less accustomed to coping with stress, who faced severe and acute stress, so-called critical incident stress. Some dealt with this better than others, in that they were able to function properly. One example is Mercury Energy's Paul de Abaitua, Manager of the CBD Supply Diversity Project and coordinator for the import and distribution of reserve power during the crisis. His reaction to the crisis and the task at hand was one that primarily showed signs of a positive stress level. Immediate reactions to positive stress, such as excitement, showing signs of being energized, alert, quick thinking, and responding with fast reactions, were some of the characteristics that featured in his work.[15] He saw the hunt for generators and competing with other actors internationally, bargaining, using contacts, etc. as rather stimulating and exciting. He displayed nevertheless, relief when the crisis was over, satisfaction with the work that he performed for the company, and confidence in the company's future ability to cope with importing reserve power (cf. Flin, 1996:104).

Mercury also took various measures to maximize the functionality of the work place at Mercury, providing room and resources, and tried to reduce stress when- and wherever possible. One of the very first things Mercury did was to provide all parties with a working space from which they could keep in contact with their colleagues. The rooms had separate phone lines, message boards, and desks. This facilitated information sharing (Cheng, interview, 11/16/98). Mercury also came to provide work areas for visiting politicians (MoCD, 1998:21).

The power company also handled a number of other practical issues smoothly. Designated parking spaces in the Mercury parking lot were set up for each actor and paper arrows were placed in the hallways so that the people would know how to get to the briefing room. Rooms for resting were also arranged in the basement of the Mercury buildings. Mercury's own staff mainly used the beds in this room when there was a slowdown in the work or when people started looking too exhausted and were ordered a break. The shower rooms in the building were also open to everybody and Mercury provided towels. Further, childcare was arranged for the people that needed to stay long hours or could not get home in time to take of their children (Cheng, interview, 11/16/98).

Mercury provided hot meals to all emergency management staff working out of the Mercury office. Bowls with packages of glucose and granola bars were available in case blood sugar levels started to drop (Cheng, interview,

11/16/98). Such seemingly mundane gestures can, in fact, affect a crisis management capacity greatly.

Another very practical aspect that the management at Mercury had to deal with concerned their own control room staff, many of whom did not want to go home after a while. They became so immersed in their work and in all that was at stake that they simply did not think that they could leave their positions. These people were under severe stress. As the nerve center for the power company they played a key role in the operative work during the crisis. How people react to stress in crisis situation varies greatly. Some people have a steep stress curve and they may quickly become overburdened by working long hours and have to be removed from their duties. Other people thrive under stress and rise to the challenge. Many crisis participants have observed in retrospect that the crisis was "fun," "thrilling" or a "peak experience." People with long stress curves can last through a crisis and perform better than ever. In crises, it is not an uncommon reaction to stress that people take on an enabling sense of personal responsibility. This seems to have been the case with some of the control center staff and other management positions at Mercury. The staff at Mercury was put under additional stress as the company was faced with anonymous threats of violence. One such threat was a bomb scare directed at Mercury on March 3. During the bomb threats the whole office had to be evacuated while the Police searched through the building (Cheng, interview, 11/16/98). That same day the company also faced a firearm threat (MoCD, 1998:42).

Notwithstanding those that rose to the occasion, many people suffered severely under the pressure and showed psychological and physical signs of not coping well during and after the power outage. The fruitless attempts at reviving the cables caused the situational assessments to be changed over and over and the time pressure only increased each time the cables failed. The sense of responsibility was very strong within the company. Mercury operations center staff were reported to have shown signs of being caught in the feeling of being indispensable.[16] Sometimes staff literally had to be pushed out of the office and forced to take a break and get some rest. The negative publicity the company as a whole was receiving caused hardship for many people working for Mercury and the company climate was gravely affected by the crisis, the public inquiries and the subsequent split-up of the board.

Most of all, however, the power outage hit the company's management. The most prominent example of this was Mercury's Chief Executive Wayne Gilbert's tragic death only days before the governmental inquiry into the power outage was to be released. Gilbert suffered a massive heart attack while in his office at Mercury on July 14, 1998. Gilbert was very committed to his work and had taken great pride the company's success. According to Mercury's new chairman Jim Macauley, the cable failures had completely devastated him (*The Press—New Zealand News*, 07/15/98). In the aftermath of the crisis some other key employees left the company. Despite management's serious efforts to move on and get past this crisis, it remains a traumatic experience for the company.

4.8 SYMBOLIC POLITICS VERSUS TECHNICAL–MANAGERIAL LOGIC

Crises are inextricably connected to the social, economic, and political context in which they occur. They are manifestations of the tensions that arise when these conditions interact in undesired ways ('t Hart, 1993:39). Consequently, there is often a lot at stake for decision makers who find themselves in a crisis situation. However, there are not only potential costs to take into account but also substantial gains that can be made. Crises entail both threat and opportunity. There is a delicate balance between PR success and disaster in crises. Crises produce opportunities for dramatization. Actors can reach out to a large audience and demonstrate resolve, empathy and commitment to core values of the political system ('t Hart, 1993:36). High visibility rituals such as state funerals, critical press conferences, official visits to disaster sights and public inquiries are examples of these types of opportunities.

These symbolic dimensions of crisis communication and crisis politicisation do not always sit well with the logic of technical and managerial responses to crisis conditions. In the Auckland case this was illustrated in at least two respects.

The so-called freezer project was a high-profile management effort in the crisis. This project was unique because it was initiated from the Mayor's office and it was one of the situations where the Mayor showed a certain amount of compassion, ingenuity, and resolve. The Mayor and the Civil Defence Planning officer from the City Council thought that the decision to place a well-guarded giant freezer in the CBD for the restaurant owners to place food in was a very good idea. The Public Health authorities did not share this opinion. They had established a good working relationship with the Environmental Health unit on the City Council, yet did not however, normally work closely with either the Mayor or the Council Civil Defence unit. Part of the problem stemmed from the fact that the Civil Defence unit on the Council did not consult other Council offices that had a sound knowledge of the CBD area, as well as the food and health issues involved.

The freezer issue then became an interagency conflict when it became clear that the Public Health authorities really did not approve of the idea. For the Health Authorities the freezer presented a perfect set-up for what they termed *time-temperature abuse*. They did not want people taking things out of the freezer in the morning and then putting it back at the end of the day as that would create major health concerns.

Various factors contributed to making the freezer project quite an ineffective management effort. First of all, by the time it had been located and set up in the CBD, most of the food that local businesses had in their shops or restaurants had already been discarded or destroyed by the Health Authorities during their inspections right after the fourth cable broke. Had the freezer been located in the area earlier, it could possibly have been used for long-term storage with the Health Authorities' blessing, but this was not the case. Secondly, the Health Authorities ended up helping out with writing the rules for the freezer after they pointed out to the Mayor that the number of people who would be going in and out of the freezer would in itself, negatively affect the temperature of the freezer.

In the end, the rules that the Health Authorities had set up for the freezer were so strict that hardly anybody actually used it. This was the Health Authorities' way of circumventing the Mayor's political move, which had worked well from a public relations point of view. Underlying the conflict over the freezer was the perceived ignorance of the Mayor and the Civil Defence unit on the City Council with respect to the Public Health authorities' role and expertise concerning food issues during the crisis. Their Public Health liaison officer at the Mercury meeting (an Environmental Health officer from the Council) had not adequately conveyed this role at the meeting. The Environmental Health Officers had not been called to the meeting at all. The only reason they ended up there was because two of the managers from Public Health decided to bring them along to the first meeting (Hope, interview, 11/18/98).

The City Council also took the opportunity to take some other symbolic actions in this crisis. The Council decided to move as many of its functions as possible outside the CBD. The main City Council building was blacked out so some of the Council members had to move in any event but the Council also saw an opportunity to display empathy and proper restraint. Representatives of the Council saw the crisis as having primarily an economic impact and they therefore took the politically important decision, not to draw power from the grid and away from the businesses, but tried to relocate instead to lessen the burden. How the public and the businesses received this symbolic action is unclear. The relocation of the Council, however, took two entire days during which the Council was unable to work effectively. For their part, the health authorities regretted not being able to coordinate and to work with the Council during the first two days of the crisis when they were visiting the restaurants and food establishments in the CBD.

From a technical and managerial perspective, moving the Council merely complicated the situation. They came late into the management of the crisis. Council personnel had difficulties in identifying and locating the basic data they needed in order to make informed decisions about the management of the CBD during the outage. The fact that the Council did not end up assuming a very prominent role in the management of the acute crisis may in part be explained by the fact that they were not up to full operation until a few days into the crisis. By then, the opportunity to take charge had, in a sense, already vanished.

It is not just that symbolic actions do not necessarily reconcile with managerial goals. The reverse is also true. One example of this in the Auckland case was the Mayor's decision to relocate to Mercury Energy's headquarters. From a technical and managerial perspective this was a logical decision since the Mayor's office was on the fifteenth floor in the CBD and it became completely impossible to work from it once the power and the elevators went off. The Mayor had to keep close contact with Mercury Energy and the other actors that were also offered office space at Mercury. In that sense his choice to move his office to Mercury was quite natural as well as functional. Mercury also had the media facilities required to broadcast public statements and press conferences, which were quite important for the Mayor as the primary political leader of Auckland. The fact that the Mayor moved his office to Mercury Energy's facilities and held his press conferences there made for some terrible symbolism, however. The Mayor became intricately connected with Mercury, especially after he had held his press conferences with the Mercury logo showing in the background. The fact that the Mayor's office was moved to Mercury suggested a sort of dependency on Mercury, which made the people suspicious that the Mayor was taking their side. The fact that the Mayor also had decided not to criticize the power company publicly until the crisis was over worked only to confirm this suspected bias.[17]

4.9 EXPERTS AND DECISION MAKERS

Expert advice is commonly enlisted in crises to inform decision making.[18] The degree to which expertise is relied upon and the specificity and degree of knowledge relied upon is however, variable. Decision makers may content themselves with expertise present in affected organizations or they may, in what appears to be a growing norm, seek expertise outside of the organization as well. Factors such as the degree of politicization as well as intensity of the event often play a part in determination of how much external expertise is deemed appropriate. Perception of acceptability typically plays as significant a role in crises as does substantive complexity (George, 1980; Farnham, 1997; Tetlock, 1992). The differing contentions between "experts" frequently observed in conjunction with high-profile events highlights the uncertainty which tends to pervade crises milieus in particular (Rosenthal, et al., 1989; Sundelius, Stern, and Bynander, 1997). Uncertainties and the plurality and disputation of "truths," make a continual parenthesis of the term *expert* sensible, as does the assumption of subjectivity in even the apparently most concrete of situations. In the same vein, classification of expertise in exclusive technical or politico-administrative terms is a virtual impossibility. Both the type and origin of expertise that predominate in a crisis both exert a strong influence on the development of the crisis management process however. The availability and seeking of expertise either within or outside of involved or stakeholding organizations has for example, an impact on this process (Rosenthal and 't Hart, 1991).

During the Auckland power outage, expertise was available and sought both within and outside of affected organizations. Different and often fuzzily-bounded forms of expertise, i.e. technical and sociopolitical, were relevant. The actor at the center of the crisis, the provider of electricity, Mercury Energy, had substantial internal expertise at their disposal that was effectively utilized during the crisis. Nevertheless, changes in the company's philosophy had resulted in significant changes to their mode of operation and organisational structure three years prior to the power crisis. These changes involved the laying-off or early retirement of staff, some of whose expertise would have been valuable during the crisis. Technical expertise which had previously been present within the organization, thus had to be sought externally by Mercury during the power crisis from New Zealand and other countries such as Australia and Canada (Cheng, interview, 11/16/98).

The government of New Zealand was also unable to provide engineering expertise. Under a program of radical economic reform begun in the early 1980s, the ministry under which such expertise was housed, the Ministry of Works, was disabled.[19] The national government was for the most part, mainly a symbolic actor during the crisis. The Ministry of Commerce conducted a post-crisis inquiry into the outage for which it relied to a large extent on external expertise. The local government relied upon its own political expertise but with regard to technical matters, almost exclusively upon the expertise so readily proffered by Mercury Energy.[20]

The Manager of Network Customer Services at Mercury Energy, Raymond Cheng, who was the manager "in charge of power restoration during the CBD Power Crisis" (Lifelines, 1998), brought to bear his previous experience in crisis management. He was able to utilize this experience during the power crisis given his position and responsibilities in the company. Cheng is educated in both engineering and business studies and has twenty-five years of experience in the electricity industry, gained in New Zealand and elsewhere (ibid.) In addition to his extensive experience as an engineer, Cheng is a registered consultant for the World Bank in the Asian Development Bank and has also worked in law enforcement (Cheng, interview, 11/16/98). Cheng had previous direct experience in crisis management from when he coordinated and managed, together with the Department of Internal Affairs, crisis management response and relief to cyclone victims on the island of Samoa (ibid.) Cheng's expertise in crisis and personnel management enabled him to respond effectively—perhaps too effectively from the perspective of certain actors—to the Auckland power outage.

Cheng's organizational skills aided him in establishing an effective ad hoc emergency center at Mercury Energy to which local government and emergency services personnel were called to meet only about sixteen hours after the collapse of the fourth cable. Emphasising Cheng's organizational prowess is not to minimize other integral factors bearing upon Mercury Energy's central role in the crisis however. What the authorities did not have to offer was perhaps just as important as what Mercury had. As stated, Mercury Energy was the owner and operator of the supply network. It had vital information and resources on their premises that other actors did not have. Neither local nor national government had an independent engineering or expert technical group at its disposal that could counterbalance the expert monopoly which quickly formed at Mercury. Mercury's prompt provision of workable solutions from both an organizational and technical perspective was badly needed by a worried and inadequately equipped government. As Rosenthal and 't Hart indicate, uncertainty and time pressure in crisis situations can result in powerful ad hoc conglomerations which are legitimated because—at least initially—much-needed expertise is perceived to be centered there.

> The felt need to quickly reduce crisis-related uncertainties heightens the need for relevant expertise and advice, often of a highly specialist nature. In crisis situations, there is the distinct possibility that because of this, experts advance into decision-making roles. Although this role expansion is made possible by the prevalence of improvisation and the blurring of formal jurisdictions during crises, a more important factor lies in the shortage of expertise in hectic moments when decision makers must make prompt choices even though they lack crucial facts and figures. Under such circumstances, those few experts who are

> able to give advice immediately may be elevated to positions of near monopolistic power (Rosenthal and 't Hart, 1991:363).

It is not difficult to understand why Mercury was relied on for expertise and resources that the company had. What is less clear is Mercury's independence in securing outside expertise. When outside expertise was required by Mercury Energy, which the company was in some instances forced to seek out-of-country when needed consultants were unavailable or nonexistent locally, Mercury Energy was entrusted to recruit this expertise themselves. The fact that Mercury Energy was a corporate actor and owner of the network explains the government's holding them financially responsible for obtaining experts when needed. It does not however, explain the city government's complete submission to Mercury Energy when it came to soliciting expertise. As the crisis in Auckland played out, the local government found itself increasingly compromised in the media and public eye due to their strong association and reliance on Mercury Energy.

The response to the power outage in Auckland lends support to Fischer's theory of the technocorporate state (Fischer, 1990). *Technocorporatism* examines "the thought systems and techniques that both guide and legitimate the practices of the technocratic decision makers, or technostructure, of this emerging state formation" (ibid.:181). In such a system, technical-managerial reasoning takes precedence over other forms and has decisive influence over decision making. Fischer notes:

> In essence, such strategies are designed to supplant political bargaining and compromise with technically rational decision-making procedures, or what has been called "methodological decision making." The task, in the ideal, is to translate—or redefine—political problems into administrative issues amenable to technical solutions. Political decisions are thus to be made on the basis of technical calculation rather than "untutored" value preferences (ibid.:182–183).

In Auckland, whether or not the transfer of ownership of Mercury Energy from the formerly publicly-controlled AEPB and the subsequent further slide of the company and its control over Auckland's infrastructure away from the scrutiny, control and understanding of elected officials can be said to epitomise such an "ideal" is debatable.

Further, it should be noted that expertise, even within technical communities, is not uniform and more often than not implies contested understandings of problems. Notwithstanding, the fact that Mayor Les Mills was reduced to likening the electricity grid in Auckland to a bowl of spaghetti when Mercury expertise was not present to provide clarification of the situation at one of the crisis meetings, is illustrative of the dependence on the company and Mercury's virtual expert monopoly. Moreover, Mercury was accorded the privilege of input in the selection of outside expertise to evaluate the crisis and was subsequently delegated the task of independent implementation of self-corrective procedures by government in the Ministerial Inquiry which investigated the outage. This self-policing indicates that consensus was strong in the Auckland case.

To sum up, the Auckland city government allowed Mercury Energy to enjoy an expert monopoly for most of the power crisis. This deprived the government of a basis for independent assessment of Mercury's claims.

4.10 CRISIS MANAGEMENT IN AN URBAN SETTING

The crisis which ensued when the electricity grid supplying central Auckland collapsed and it was recognised that virtually the entire CBD would be without power for a considerable period of time was defined by special features due to its urban character. The special features of what have been referred to as "urban crises" are distinguishable and significant from both cause and effect perspectives.

While there are certain common denominators in urban crises, a further distinction has been made between those occurring in industrialising or "developing" countries and urban crises in wealthier, post-industrial nations (Jacobs, 1997). As Jacobs points out, "[i]n developed nations, the emphasis is upon consequences of decline in old cities; de-industrialization; the degradation of the environment; and the dilapidation of old physical infrastructures" (Jacobs, 1997:127).[21] While the Auckland crisis had other distinguishing and intervening features such as demographic change involving a steady influx of inhabitants to the region and thus, a growing burden on infrastructure generally, the age and condition of the power grid supplying the city was indeed a factor.

Urban crises are indeed usually tied to "complex and risky preconditions" (Jacobs, 1997:127) as cities such as Auckland are places of high traffic supported by complicated infrastructures, systems and routines. Cumulative problems in urban environments

> provide the conditions for the development of generalized and sectoral emergencies.[22] Where problems cluster to threaten a breakdown, city officials have to deal with a multiplicity of urgent and threatening situations either together or within a narrow time-span (Jacobs, 1997:128).

This was certainly the case in Auckland where an actor in the electricity sector—Mercury Energy—was ultimately struck with a breakdown which city authorities and emergency services had to manage under time pressure. The most immediate effects were distinctly urban in nature. The fire department knew that there was some sort of problem of significance prior to receiving notification of the grid failure, as soon after the final cable went, they began receiving a myriad of calls reporting individuals trapped in elevators and car parks and fire alarms going off all over the city (the latter due to backup batteries being dead) (Warby, interview, 11/17/98). Some doors to buildings, like the elevators, were electrically operated and when the power went off they were fixed in either a closed/locked position or left open/unsecured (ibid.)

Once people began operating generators on a large scale, problems occurred, mainly with respect to ventilation and fire risks. The operation of large numbers of generators on the city streets posed ventilation, congestion, traffic, and noise concerns, as well as the general fire and safety hazards entailed in operation of such diesel-fired heavy equipment. The fire risk and pollution produced through the large-scale operation of generators in Auckland over a long period of time is not difficult to imagine when one considers that the diesel-operated machines require refuelling approximately every three hours (de Abaitua, interview, 11/16/98). Some problems with foreign generators occurred as they operated on different voltages; these problems were easily solved where conversion cables were available, used and hooked up properly, but posed danger and expense in a number of cases where they were not (ibid.) Emergency services also responded to complaints of generators being operated in closed and poorly-ventilated locations, such as the stairwells of buildings where generators were placed in an effort to power offices (Warby, interview, 11/17/98). Such activities posed fire risks as well as the risk for asphyxiation. A further problem with generators reflects both the urban and the economic nature of the crisis. Due to the initial high demand for generators which eased gradually with incoming arrivals, price gauging as well as what de Abaitua referred to as a "black-market for generators" emerged, in which less ethically-inclined opportunists profited from charging exorbitant rates and providing sub-standard products (de Abaitua, interview, 11/16/98).

Among the specifically urban-related problems which became apparent as the crisis progressed were difficulties with high-rise buildings. There were a number of problems with tall buildings during the crisis. With a limited supply of electricity, it was difficult to pump fresh water up past several floors. This led to expressions of concern over the possible outbreak of Legionnaires Disease due to consumption of stagnant water, a concern which thankfully never materialised (Hope, interview, 11/18/98; Pritchards, interview, 11/18/98). While the residential population of central Auckland is very small for a city of its size, for those who did live in tall buildings in the area, walking up numerous flights of stairs was exhausting, particularly as the outage took place during a summer heat wave and the weather was extremely warm. Built of concrete, the tall buildings absorbed and retained heat such that they were horribly hot and uncomfortable, a factor stressed by the city fire chief (Warby, interview, 11/17/98).

There are an increasing number of similar findings regarding tall concrete buildings in the study of heat buildup in cities, a relatively new and expanding field of research. Densely populated and built-up areas are particularly susceptible to developing what are referred to as "urban heat islands," "bubble-like accumulations of hot air [which] develop over cities as naturally vegetated surfaces are replaced with asphalt, concrete, rooftops and other man-made materials" (Steitz and Tyson, 1998:1). These heat islands form during the hot summer months when already high temperatures are driven even higher by the shape and form of urban space. NASA studies of American cities have revealed that heat buildup in cities is particularly dangerous because it allows ozone to accumulate at ground level, the inhalation of which causes serious respiratory irritation (ibid.) While it is in no way certain that urban heat islands developed over Auckland during the crisis as no documentation has been found regarding the phenomenon in the area, it is clear that heat built-up heavily in the Auckland center during this period, a period in which there was also very little precipitation.

The extremely hot and dry ground conditions in the Auckland CBD have, in fact, been held up as a likely factor contributing to the instability and ultimate collapse of the electrical grid, particularly by the media. While the ministerial inquiry into the outage indicates that ground temperatures in Auckland were grossly underestimated—about seven to nine degrees lower than the actual February ground temperature of 22 to 24° C—the inquiry concluded that "this temperature difference is of little or no significance in respect of the failures which occurred" (Ministry of Commerce, New Zealand, 1998:65). The dryness of the soil however (exacerbated by the heat) appears to have been a greater factor in the failures because this aridity affected heat dissipation from the cables (ibid.) During the crisis, Mercury Energy enlisted first BICC (the Penrose-Quay gas cable manufacturer), and following their inconsistent and high test results, further consultants from the companies Geotherm and Worley to test the soil conditions in the CBD. The second group of tests suggested that a thermal resistivity[23] of "between 2.50°C m/W and 4.0°C m/W for the sand backfill should be assumed in assigning cable ratings" (Ministry of Commerce, New Zealand, 1998:65). When the cables were installed however, a thermal resistivity of 1.2°C m/W was assumed, a rating which was obviously too low and was not subsequently adjusted as it should have been (ibid.) The thermal resistivity rating is important as it is tied to maintenance of the conductor temperature under the "worst-case" scenario temperature of 85° C (ibid.) According to Geotherm, "even though thermal stability analysis was not undertaken, it was highly likely that the sand backfill was thermally unstable and prone to thermal runaway even under 'normal' loads" (ibid.)

The findings documented in the Ministry's inquiry with respect to soil conditions and temperature clearly raise as many questions of an organisational and political nature as they do technical and climatological. The complex character of crisis is evident in the Auckland example. One such complexity is the relationship between urban crises and privatization, which is pertinent to discussion of the Auckland power outage as a sectoral urban crisis. Jacobs observes,

> [a] crucial issue is that of the effectiveness of market policies in overcoming urban problems. The promotion of market policies provides a theme, internationally, because state-funded initiatives in world cities continue to lose ground in favour of privitization, contracted local services and the establishment of public/private partnerships. The market context influences both broader city-wide and regional policy strategies and the practical operational arrangements designed to deal with sectoral or localized problems (Jacobs, 1997:128).

Jacobs' comments are applicable to the Auckland case and raise interesting questions with regard to the public/private interface, key public infrastructure and regulation and the character of cities as economic loci. As detailed earlier in this study, New Zealand has undergone extensive privatization and corporatization in recent years, reform which has had an extensive effect on all sectors of the economy and the polity, including the electricity sector and local government. The Auckland Electric Power Board (AEPD) as a semipublic body had a stronger connection to the polity and at least theoretically, greater accountability to the polity. Crisis researchers point out that cities "should indeed be considered to be disaster-prone to the extent that infrastructural arrangements of urban society often represent high-risk technologies that are definitely not failure-free" (Rosenthal et al., 1994:116). A very reasonable conclusion to draw from this observation is that the high-risk technologies and infrastructures of urban society should be carefully safeguarded and the actors maintaining them responsible and accountable, be they public or private.

The urban core of Auckland reflects an emphasis on private capital common to prosperous urban cores. The economic consequences of the Auckland power outage for the city and the country as a whole were enormous and illustrate the complex impact of urban crises. The crisis was basically economic and political as serious risks to health were limited and well-contained. The city of Auckland, like other major cities, is a center for business. While not the capital, the city is New Zealand's largest, has a much greater representation of branch offices than anywhere else in the country and is an important international port and business junction in Australasia. Like other comparable cities worldwide, a disproportional contribution to the nation's Gross National Product (GNP) is generated from Auckland. Businesses within the Auckland CBD suffered economic losses of varying degree depending on the size and nature of the business.

However, the Auckland power crisis also reflects the resourcefulness of cities. While small businesses did not benefit from the crisis—thousands of small businesses sustained severe financial losses, many of them forced into receivership or bankruptcy—bigger and wealthier businesses were able to innovate and, utilizing information and communication technologies, conducted their business affairs from locations outside the Auckland CBD. Some businesses resumed operation from locations as far away as Wellington or cities outside the country, such as Sydney, Australia. According to a survey of 129 organizations in the Auckland area, "representing 6,450 employees or more than 10% of the CBD workforce," conducted by property consultant Colliers Jardine directly following the power crisis, "79% of the respondents moved at least some staff from their office, but only 6% of respondents moved to a pre-arranged alternative office as part of a contingency plan" (Sheehan, 1998:7). Approximately one-quarter of those surveyed stated that the crisis had led them to consider permanent relocation options as well as mobile offices and home-work (ibid.)

There are some other advantages to a crisis occurring in an urban setting which were observable in the Auckland case as well. Services and personnel are more readily available in urban centers. Although a certain amount of expertise needed to be recruited in the Auckland power outage, this imported expertise was of a technical and engineering nature while most of the recovery operations were accommodated by emergency services and other officials at the site of origin. This benefit has been noted in other urban crises, such as the Bijlmer air disaster of 1992 in Amsterdam. The need to involve other levels of government in the response phase was not there as the disaster "happened in a major capital city—Amsterdam—which has at its disposal large-scale, professional and specialist resources" (Rosenthal et al., 1994:101).

An important reflection on urban crises based on the Auckland experience is instructive. When a crisis occurs in a major urban center which is the economic locus of an area, or as in the case of Auckland, the entire country, special attention should be directed towards the importance of international perception and portrayal of the event. The international media depicted the crisis in Auckland as being more severe and encompassing than it actually was. Sensationalized news releases have been blamed for adversely affecting tourism and travel to Auckland during the crisis, traffic upon which Auckland's businesses are economically dependent. Shortly after the collapse of the fourth cable, BBC world news advised in a story dramatically entitled "Auckland's Heart of Darkness:" "Don't go looking for the bright lights in Auckland, New Zealand—for the next week, the only electricity in the centre of the country's biggest city is likely to be in hospitals and other emergency services" (BBC News, 1998).

4.11 SEQUENCING AND SYNCHRONICITY

Sequencing and synchronicity are useful analytical themes for studying crises such as the Auckland power outage as they help to sort out the complexity of decision-making problems which decision makers are confronted with in crises (Stern and Nohrstedt, 1999). Sequencing is where the crisis is viewed "as a cluster of interrelated decision problems where prior choices enable or constrain subsequent ones in significant ways," whereas synchronicity refers to the occurrence of parallel or coincident events in a crisis which may or may not be significantly related to events, but may exert an impact on decision-making processes nonetheless (ibid.:310). Examples of both sequencing and synchronicity are found in the Auckland case. Noteworthy examples of synchronicity are found in the accidental clipping of cables in the grid two days prior to the first cable failure, in the electricity crisis of a similar nature which took place in Queensland, Australia during the Auckland power crisis, and in the second escalation of conflict in the Persian Gulf which also broke out during the Auckland crisis. Sequencing in which choices made early on in the crisis constrained and progressively limited later options are observable in both the initial optimistic estimate regarding when the cables could be repaired and the power supply grid would be up and running again (it turned out that the damage was in fact, largely irreparable) and in the initial decision to advise people to stay out of the city center.

A contractor working at Liverpool Station on January 20 accidentally clipped controller cables which were attached to the Mt. Roskill station. This resulted in an immediate disconnection of the No. 1 110 kV oil cable between Roskill and Liverpool, as well as the overloading and subsequent disconnection of the parallel cable, the No. 2 Roskill-Liverpool oil-filled cable. As mentioned in section 6.1, these disconnections were short-lived, but they did result in increased loading on the Penrose-Quay line—a load that, while ascertained as within the acceptable rating, appears to have increased grid vulnerability. Two days later, the first cable, the No. 2 gas cable at Penrose-Quay suffered a complete failure. Thus, while the accidental clipping of the cable appears to have had a decided impact on reducing system reliability, the event was, at least not initially perceived by operators at Mercury Energy as problematic. The system was seen to be resilient enough to tolerate such an incident and remain within normal operating limits; therefore, reliance on the system in a "business-as-usual" fashion continued—up until the complete collapse of the first cable two days later.

Two international events coincided with the cable failures in Auckland which rendered response to the crisis considerably more difficult. An electricity supply crisis also broke out in Queensland, Australia during the Auckland power outage and the escalation of conflict in the Persian Gulf broke out during the same period. Because the crisis in Queensland, which began with rolling blackouts on February 24, was of the same nature as the Auckland crisis and occurred at the same time, it meant that expertise required in Auckland was already committed in Australia. The conflict in the Gulf had the effect that heavy-lift aircraft—always in rather high demand due to their scarcity—were more difficult to get hold of than usual.

A further slightly ironic example of synchronicity was that of a Mercury-sponsored concert which was planned to take place on the February 28 in Auckland. The unfortunate correspondence of the "Schweppes Showtime" summer concert with the electricity outage did nothing to improve Mercury's failing image, particularly given the symbolic impact of the concert advertisement. Co-sponsor of the event, Mercury Energy, had committed themselves to providing the lighting for the concert (Gardiner, 02/24/98). The advertisement reading "Mercury Energy looks forward to lighting up the lives of hundreds of thousands of Aucklanders this Saturday" was mailed out to Auckland residents the week the fourth cable broke (Gillon cited in Hansard, 24/02/98). Cancelling as-yet-unreleased advertising and materials for the promotion-turned-fiasco, Mercury appeared anxious to renege on their commitment to the concert. Mercury staff were "worried whether [their involvement in the concert would] send an inappropriate message during the crisis" (Gardiner, 02/24/98).

With respect to sequencing, there were several significant chains of events in the case where it is evident that initial decisions taken exercised impact on the availability of options later on in the crisis. In this way, possibilities were progressively funneled into a narrower range circumscribed by actions or nonactions early on. This pattern is evident in Mercury Energy's initially optimistic prognosis of the cable failures, which led to a slower start in completing the overhead line. Once it was realised that the cables could not be reconditioned as hoped—indeed, as Mercury employees had basically calculated—great pressure was felt to get the overhead line operating as soon as possible.

The sequence which began with Auckland Mayor Les Mills' broadcast appeal to the public to stay out of the CBD while the power was out is another example of sequencing in which this first choice had observable consequences for those which came later. City Council actually needed to concoct schemes and plans such as the "Shop 'till you Drop" contest (discussed in the previous chapter) in order to entice people back into the city once it was realized that problems of criminality and chaos were not materializing as anticipated. Thus these sequences usefully illustrate the "path dependency" typical of crises.

4.12 LEARNING

One of the problems that arises when talking of organizational learning is an ontological one. Who or what in a collective unit, such as an organization, can be said to learn? At the bottom of this is the issue of whether only individuals can learn or if there can be some sort of systemic learning in larger social aggregates. The literature supporting the idea of learning from an institutional perspective often focuses on experience-based learning, where the development and change of roles, rules and routines based on experiences of social problems is what constitutes the actual learning process. According to this perspective, however, organizations and institutions not only learn but can also forget if institutional memory is not cultivated and made accessible to the actors involved. Another interesting question linked to the possibility of experience-based learning is what possibilities of learning from the experience of other organizations exist (Stern, 1997:50). This can prove to be quite a cost-effective alternative to learning from one's own experience and it is one of the reasons why it is interesting to analyze the concept of learning and to identify lessons learned from various crises.

It may be useful to divide learning into three categories. Firstly, precrisis learning can be said to include prior personal experience and the more or less sophisticated use of history by policy makers. Secondly, an intracrisis learning phase concerns the lessons that actors have learned within the span of the crisis. This type of learning is often part of an acute coping mechanism and/or takes the form of innovative solutions that often lead to the development of new procedures. Thirdly, post-crisis learning is often driven by public inquires and other investigative ventures. Likewise it includes drawing upon the lessons learned from the crisis, implementing changes that have been identified as needed, and transferring the knowledge gained from one crisis to future crises. This study on the Auckland power outage is, in a broad sense, part of post-crisis learning as it attempts not only to contribute to the science of crisis management at large but also aims at generating knowledge for policy makers who will need to deal with future crises.

Learning has a lot to do with the problems related to the way the human mind relies on heuristics in order to give meaning to the world and the events around it. In cognitive psychology the term *heuristics* describes the various processes and short cuts the mind uses to make sense of or as a way of finding structure in situations that are unfamiliar to it (Khong, 1992; Nisbett and Ross, 1980).

Of the most common types of heuristics people use in crisis situations are "historical analogies" (Khong, 1992). Historical analogies are used as knowledge structures in information processing and for the comprehension of new situations Khong points out that this analogical reasoning often reaches further than the actual information available justifies. Khong, whose research focuses on the bad use of historical analogies, states that the characteristics of the analogy picked are often accepted uncritically and are used to fill the information gaps that exist in a new situation. One of the problems associated with this type of heuristic thinking is that persons who are in crisis situations conspicuously often pick one analogy (often the first that springs to mind) without thoroughly examining the true parallels and differences between the two situations. Another part of this problem, Khong argues, is that analogies often are used as a substitute for proof with sometimes dire and unnecessary consequences for the decision-making process (Khong, 1992:29–30).

Historical analogies were not as prominent as one might think in this case. There had not been a similar power outage in New Zealand before. There had been shorter outages in Auckland on and off over quite a long period of time but nothing as extensive as this crisis. There were other cases, however, of major power outages and blackouts both in urban and rural settings that had taken place in other countries that could have been employed as reference cases (for example, the New York City blackout, 1977, and the ice storm in Eastern Canada, 1998).

Among the participants in the Auckland power outage, there was some mention made of the events that took place in the New York blackout. The participants that explicitly referred to this historical case were the police. The police knew of the riots and the destruction that had taken place during this event, and seemed to have based their initial response on the likelihood of something like this taking place in the Auckland CBD. To get more specific information on the New York incident and possibly obtain some tips on how to prevent this type of development, the Auckland Police searched the Internet and contacted colleagues in the United States via their computer-based networks and e-mail. These contacts seemed to have generated little substantial advice but the Police still used the analogy when setting up their initial response plan. The police saturated the CBD with officers and dispatched a surveillance helicopter to circle the CBD in order to deter potential offenders, and thus prevent the crime wave they feared. Some time into the crisis, the Auckland Police recognized that the analogy did not fit and highlighted the differences with the American experiences that they thought had contributed to a peaceful unfolding of events in Auckland.

There are a number of examples of intracrisis learning in the Auckland power outage. The Public Health authorities served as one good example of an organization that took on the new experiences during this crisis and consequently learned substantially throughout the duration of the crisis. In order to ensure the safety of people living in the area, Public Health had to reach the people in the CBD as quickly as possible—a lot sooner in fact than many other agencies. Potential problems associated with sewage and water pumps had to be checked quickly and the health authorities needed somehow to visit four hundred residential buildings. The Health Authorities divided up the territory into various teams and managed to reach all the buildings in a few days.

Another learning item was that the doors of many buildings had locked automatically and that led the health authorities to work closely with building inspectors. Up until the Auckland power outage Public Health had not dealt with food damage the magnitude that was displayed in the crisis either. They were used to having contact with restaurant and storeowners through the periodic health inspections for which they were responsible. The health inspection team, however, had never encountered a situation in which these business owners were facing many thousands of dollars in losses because of rotting food. They did not have an emergency inspection plan, which incorporated an area as large as the CBD. Scenarios had been constructed on the basis of a potential power outage, but the simulated plans never incorporated an area as big as the CBD and thus did not take into account the amount of damage that actually took place.

The health authorities knew that there was a great incentive for business owners to save food in any way possible even when it was not safe to do so. They developed a routine during this crisis whereby they distributed information leaflets upon their first visit. They then came back and inspected the actions taken during the second visit, and at the third visit they convinced business owners using at times, quite radical measures. Pouring bleach onto spoiled food products was an innovative solution aimed at avoiding the distribution of bad food as well as terminating arguments with business owners about whether or not the food could possibly be used. The bleach constituted a solution that likely took the owners by surprise but it was both convincing and effective. The fact that the health authorities were forced to work in a manner in which they never had before helped them learn a lot, especially during the crisis itself.

Various inquires which have been produced on different aspects of the Auckland power outage provide examples of attempts at post-crisis learning. The kinds of lessons learned and the general conclusions of such inquiries facilitate learning for others outside the group of people that were directly involved in handling this crisis. This section concludes with a brief description of two of the most important inquiries produced within New Zealand following the outage, the Ministry of Commerce's inquiry into the Auckland outage, *Auckland Power Supply Failure 1998: The Report of the Ministerial Inquiry into the Auckland Power Failure* and the Ministry of Civil Defence (MoCD) report entitled, *The Auckland Electricity Supply Disruption 1998: Review of Emergency Management Response*.[24]

The most significant inquiry into the Auckland power crisis was that published by the Ministry of Commerce, the Auckland Power Supply Failure 1998: The Report of the Ministerial Inquiry into the Auckland Power Failure, a nearly two-hundred page document published by the central government in July 1998. Toft and Reynolds (1994) have observed that "[t]he definition of the terms of reference of an inquiry is key to its outcome" (Meertens, 2000). Of course, the degree to which the terms of reference are followed in an inquiry and how the specified terms are formulated influences the outcome. The Terms of Reference for the Ministerial Inquiry into the Auckland outage (footnoted in chapter 6) were comprehensive, and pitfalls—such as determination of "liability, breach of law, or compensation" which have been found to lead to blaming and detract from learning—were explicitly avoided (Ministry of Commerce, New Zealand, 1998; Meertens, 2000). However, this very advantage can be seen as somewhat of a liability as well as an asset. A problem common to governmental inquiries is that they often have no binding power to ensure that alternative courses of action, which would provide for better operating practices in the future, are integrated into mandates and daily operations. As Toft and Reynolds (1994) point out, "In order that a particular type of incident does not recur it is important that the recommendation become part of the changing fabric of organisational life" (Meertens, 2000:5).

The findings and recommendations of an inquiry are an important way of summing up some of the problems that were uncovered as they suggest different actions and modes of operation for the future. The suggestion of alternative actions illuminates the most urgent problems which need to be addressed. The Ministerial Inquiry into the Auckland power outage ends their report with a list of "Conclusions and Recommendations" which include thirteen recommendations—the first eight of them specifically directed at Mercury Energy (Ministry of Commerce, New Zealand, 1998). The inquiry, first noting that the company's "response to the power supply failure, both in the immediate emergency response and in the subsequent investigation, has been of a high standard and should be formally acknowledged" (ibid.:164), goes on to recommend that:

> 1. Mercury Energy establish a specific management plan for its 110 kV CBD transmission, recognising that it is different from its distribution assets; 2. Mercury Energy periodically have its 110 kV cable management plan peer-reviewed by external experts in 100 kV cables of similar design and age; 3. Mercury Energy review its practises in respect of its use of external contract services to establish clear and definitive specifications for contract services; 4. Mercury Energy review its strategic plan to ensure that the core business of distribution and security of supply is given appropriate priority and allocation of resources; 5. Mercury Energy review its risk management processes so that all risk is managed within one systemic process to a defined standard; 6. Mercury Energy obtain, and publish, an independent expert validation of its current and committed plans for restoration of CBD security of supply as it will be by 1 December 1998 and also as it will be upon completion of the tunnel project; 7. Mercury Energy institute periodic technical audits of all major assets associated with the CBD power supply; 8. Mercury Energy review its network services contracts for each class of customer and revise them to: specify defined reliability standards for all outage times; and specify the corresponding liability for each outage period; in order that Mercury Energy has incentives to manage security as a commercial risk and enable their customers to see and understand their risk exposure" (ibid.:164–165).

Recommendations nine and ten are directed towards Mercury Energy and the AECT, the former suggesting that the company and the Trust "maintain and enhance the cooperation established in May and June 1998 so that effective corporate governance is achieved" and the latter that "Mercury Energy negotiate a Statement of Corporate Intent [for more on Statements of Corporate Intent and the interesting exclusion of Mercury Energy from this standard practice see chapters 2 and 4] no later than 1 October 1998" (Ministry of Commerce, New Zealand, 1998:165–166). The three final recommendations in the inquiry were directed at the government:

> 11. The Government consider (and determine) whether the establishment plans of network companies have been completely carried into effect and can be formally accepted as expired. If the Government considers that establishment plans are not expired then the Government should consider, in consultation with Mercury and the Trust, whether, in the public interest, steps are necessary to restore the intended

> lines of accountability to normal standards in Mercury Energy's corporate structure; 12. The Government consider incorporating, in the electricity power company disclosure regulations, a requirement for every network operator to publish, every three years, its asset management plan together with past and future security standards for each consumer class, and supply area and each distinct voltage level; and 13. The Government encourage and, if necessary, facilitate the electricity distribution industry to develop customer contracts for network service providers that reflect security of supply standards and liability provisions appropriate to each type of consumer and which give network companies incentives to manage security on a commercial basis (ibid.:166).

Striking in the recommendations for Mercury Energy is the degree of self-supervision inherent in the suggested courses of action.

Another important inquiry produced in the wake of the Auckland outage is the Ministry of Civil Defence (MoCD) report entitled *The Auckland Electricity Supply Disruption 1998: Review of Emergency Management Response*, submitted by the Director of Civil Defence, John Norton to the Minister of Civil Defence on June 18, 1998. This aim of this report was "not intended to cut across the work of the Ministerial Inquiry [of the Ministry of Commerce] set up to look at the event itself" (MoCD, 1998:1). Rather, the MoCD report was explicitly a "report on the emergency management implications of the Auckland power outage, with emphasis on lessons for civil defence organisations" (ibid.:1). In an attempt to "identify those things which went well and those which turned out less well" (ibid.), the document posed the following six questions:

> Q.1: Was there a need to declare a state of civil defence emergency? Factors to be considered include—the initial assessment of the scale of the problem; the information available to decision-makers; subsequent assessments and decisions made. Q.2: Was there an effective response to the crisis by the key responders? Organisations include Auckland City Council, emergency services and other agencies (e.g. Ministry of Health). Q.3: How adequate was the public information process? Factors to be considered include—agencies releasing information; media response to agency releases; public notices; other methods. Q.4: What was learned about the availability of portable generation to substitute for grid electricity? Factors to be considered include—availability of generators in New Zealand and overseas; shipping power generation; any other options. In contingency planning for a catastrophic emergency, is there a need for a national database of power generating equipment? If so, who should be responsible for its compilation and maintenance? Q.5: Are there other critical supply or provision issues which were, or could have been, affected by the crisis and which may be applicable to other local authorities? Q.6: What was the contribution made by Ministry of Civil Defence staff. Did it add value? (ibid.)

The forty-two page MoCD report analyzed the preceding questions and arrived at the following conclusions with respect to them:

> Q.1: In view of the requirements of the Civil Defence Act, the decision not to declare an emergency was the correct. Q.2: Insofar as no lives were lost or personal injuries sustained, the response of all those involved must be commended as effective. To what degree distress or plain stress could have been reduced will depend on individual assessment. There is a need for agencies to agree which of them will be the coordinator during events which require a multi-agency response, but fall short of a declared state of civil defence emergency, and where some focal point is clearly needed. We will also make some other suggestions. Q.3: The public information process was adequate to the extent that no one was injured directly as a result of the event, or through some incident brought about by carelessness or folly. No one was poisoned by bad food or by carbon monoxide, no one set fire to their flat by unwise improvisation. The safety and health messages seem to have been timely, accurate and effective. At the same time, there are pointers for better future results. Q.4: There are numerous companies in New Zealand able to hire out generators of different sizes. Some companies provide good refuelling and maintenance services as well if needed. They are also able to source generators overseas. There is no pressing need for a national inventory. At the same time there would be an advantage in local civil defence organisations and electricity companies considering making priority lists of those who would need generators. There are some factors to keep in mind. Q.5: The event reaffirms the community's heavy dependence on reticulated electricity. It shows the value of having other energy sources where those are practicable, and of having a back-up to line electricity. A number of organisations found that their own generators either did not perform as expected or had not been designed for the long event. Emergency management agencies, others involved in public safety, the Media, and any who have statutory duties or other compelling reasons to continue operating through a long power cut, should review their electricity arrangements. Q.6: The Ministry of Civil Defence contributed to the event response at several levels. While the Director provided advice and information to Ministers and key Central Government agencies, field staff monitored the event, offered information and advice, provided a conduit to the Central Government and ensured a flow of timely factual material in their Sitreps. Their work was appreciated by Central and local government (ibid.)

The mandate of the Ministry of Civil Defence report was to broadly "indicate what had gone well and what less well" (ibid.) MoCD staff enlisted for this overall evaluation concluded that the following were perceived as well-managed during the outage: "Inter-agency cooperation and communication; Good public cooperation and acceptance of advice; Central Government agencies were appropriately involved; Immediate public needs were identified and addressed; Other Lifeline capabilities were assessed and managed appropriately" (ibid.:2). With respect to activities during the crisis which the MoCD regarded as necessitating improvement, the report noted the following:

> Response agencies' understanding of each others' roles and limitations in events requiring a multi-agency response, but not warranting a declaration of emergency, and a mechanism by which leadership is established and promulgated in such cases; An understanding of the different communications priorities of emergency managers and the Media; Recognition of key information needs for diverse community; Business continuity planning for the specifics of a Lifelines loss or interruption; Care in the content and balance of media releases containing forecasts of future resolution (ibid.)

A prominent member of Australia's civil defence community, Deputy Director of Victoria State Emergency Service (VICSES), Gareth Davis, traveled to Auckland when the outage struck and for four days thereafter observed response to the crisis in progress. In his preliminary report entitled Visit to Auckland to Observe Power Crisis, Davis noted that his report, due to the fact that he was in Auckland only during the initial phase of the event, could only be regarded as "a snapshot of a significant event" (Davis, 1998b:1). In his seven-page report Davis discussed the delicate issue of emergency declaration in a general sense, described the CBD of Auckland and the problem which ensued, the direct effects and initial response, as well as actions which followed the initial actions and secondary effects. He went on to describe subsequent actions and other points of interest. Davis concluded his report with recommendations for Victoria in Australia, which provides a good example of a concrete effort at vicarious learning. The first three of his nine recommendations stress the importance of building up Lifelines capacity as "an integral part of local government planning and [adds that plans] must be included in the risk management process" (ibid.:7). Davis notes that special arrangements with respect to Lifelines must be made in rural areas and suburbs of cities, which may overlap administratively with rural regions (ibid). From observing problems encountered in Auckland, Davis was able to map out areas of concern for his community in Australia and to suggest appropriate preparatory actions which would reduce negative impacts should an extensive outage occur in Victoria. He suggested the following concrete actions be implemented in Victoria:

> A registry of building managers be established to facilitate communication concerning any emergencies. These managers themselves contribute to a database

> of lift and alarm maintenance companies to assist during power failure. Major installations, including critical government departments, be encouraged to install emergency (generator) power to cover at least their health and safety issues. Businesses and residential apartment blocks be encouraged to at least pre-wire for easy installation of emergency generators. Regulations concerning residential apartments be reviewed in the areas of lighting, ingress, egress, sewerage and general health and safety in the context of the loss of power. Power networks be reviewed for the ability to isolate certain critical elements, such as traffic lights. All businesses and government departments be encouraged to address seriously the subject of "business continuity" (ibid.)

Another outside report on the Auckland outage produced from a decidedly longer distance and indicative of strong international interest in the crisis is that of Lennart Göhl and the Swedish State Energy Authority (Energimyndigheten), New Zealand's Electricity Market: Regulations Scrutinised Following the Serious Power Outage in Auckland 1998. While this brief report of fiteen pages provides primarily an overview of the crisis as it occurred in Auckland and summarizes the findings of the subsequent Ministerial Inquiry of the New Zealand Government, examples of vicarious learning are noted in both the report's foreword by Deputy Director General Håkan Heden and in the reports conclusion. Heden observes that the "so-called light regulation" in New Zealand applied in earnest to the electricity market from the beginning of the 1990s is "more a type of self-regulation" of the industry (Energimynigheten, 1999:1). The director further notes that the paralysing outage has "woken anew the debate on New Zealand's model for regulation of the electricity market and the possible shortcomings and limitations of such a model" (ibid.) The report concludes with the following statement: "The electricity outage in Auckland has, through its comprehensive and important societal consequences, re-illuminated the vulnerability of modern societies' when confronted with extensive power disturbances within the electricity sector" (ibid.:15).

While it would be extremely interesting and fruitful to analyse the recommendations of the reports, compare them to the Terms of Reference or questions posed at the beginning of them and take an inventory of what changes have been implemented in the time since the Auckland outage occurred, it is neither the objective of this study nor within the scope of this report to do so. Analysis of implementation would require a substantial research effort not possible within the current report.

Let us conclude this section by reflecting upon the nature of the inquiry process. Governmental inquiries in particular are highly complex, if short-lived institutions. The manner is which they are constructed, the questions they formulate, their operating procedure and how their conclusions are arrived at are arguably as interesting as the reasons for which they were created. One of the potential problem present in inquiries is that of bias (Lindblom, 1980). Based on their analysis, Toft and Reynolds (1994) have noted that public inquiries tend to be quite subjective in their design. They "are not always conducted on the basis of disinterested objectivity and open-minded construction. Parties involved in public inquiries have, consciously or not, their own hidden agendas. Politicians, authorities and the public may all have their paradigmatic world-view" (Meertens, 2000:2). The design of inquiries obviously has an impact on learning potential.

NOTES

1. The "security" of a power system, along with a system's "adequacy," are the two basic aspects of power system reliability assessment (Anders, 1990:270). "Adequacy relates to the existence of sufficient facilities within the system to satisfy the consumer load demand or system operational constraints; it is therefore, associated with the static conditions, which do not include sudden system disturbances. Security on the other hand, relates to the ability of the system to [adjust for] sudden disturbances arising within that system" (ibid.) The term "reliability" may be used interchangeably with "adequacy."

2. See appendices for Mercury Energy's figure depicting CBD electricity demand from 1955 to 1997.

3. For more on linear, complex and unexpected interactions in technological systems, see Charles Perrow's *Normal Accidents: Living with High-Risk Technologies*, 1999 Princeton University Press, Princeton, N.J.

4. While a power grid provides an example of a very tightly coupled and interdependent system, an example of a loosely coupled system is an assembly line, where disruption in one area is precisely that, disruptive, but which can in fact be accommodated through alternative arrangements such as delays, reorganisation of sequences and substitutions such that the impact need not be disastrous (Perrow, 1999).

5. Perrow has developed a theory of "normal accidents" (NAT) respected in both the natural and social sciences which in summation claims that "no matter how hard we try there will be serious accidents because of the interactive complexity (which allows the inevitable errors to interact in unexpected ways and defeat safety systems) and tight coupling (in which small errors propogate into major ones) of most risky systems. Catastrophic accidents are 'normal' (though rare) because they are inherent in the system (Perrow, 1994:212)."

6. See appendices for Mercury Energy's electricity demand plots on all four cables under the summer months of 1995, 1996, 1997, and 1998.

7. A similar problem-framing situation confronted the Carter administration when the Iranian Shah lost his power in 1979. During the political turbulence preceding the fall of the Shah, the only question posed in the White House for quite some time was "How could the Shah be strengthened?" Nobody posed the question, "What do we do if the Shah falls?" (Moens, 1991:211–37).

8. For more information on the access to and use of experts in this case, see *Experts and Decision Makers*.

9. For more details as to how the operations center functioned, see the section *Stress Management*.

10. For a more elaborate discussion of the media management at Mercury, see the section, *Crisis Communication and the Mass Media*.

11. For more details on the information conflict between the police and Mercury Energy, see *Intelligence Gathering and Information Processing*.

12. One example of such an issue was the Hero Parade, the annual New Zealand Gay Parade, which was to be held on February 21, 1998. There was a big debate as to whether the City Council would help pay for crowd control and cleanup after the parade. The Council was split fifty-fifty on the issue and the Mayor then voted against economic support. This made him very unpopular in the press and with some of the population.

13. According to Dallas Moore (Manager, Planning and Public Education, Ministry of Civil Defence, Wellington), the initial message left a rather negative and inaccurate impression internationally (Moore, interview, 11/20/98).

14. This response can have very adaptive and positive effects. It occurs when the brain registers something that it perceives as a threat in the close environment and signals the rest of the body to produce an energy boost that will allow the individual either to flee a threat or to fight an agressor. Any type of situation that an individual finds unusually challenging or demanding can trigger this response. In emergencies, professionals dealing with such situations often experience this effect only through exposure to traumatic scenes or chaos (Flin, 1996:98).

15. For an overview of the impact of positive and negative stress, see Flin, 1996:104.

16. This sometimes happens when people become overly engrossed in the course of events and feel very personally responsible. Eventually they cannot see themselves leaving their post or assignment, even for a limited time period. They see the emergency work as far too important for personal considerations and fear that everything will collapse if they stop doing what they are doing. They feel as if they would be "abandoning the ship" were they to leave and therefore disregard their need for rest and food, necessary for functioning. The risk for burnout in these situations is substantial and by not listening to one's own needs in this manner, an individual responding in such a way may in fact exacerbate or complicate an already difficult situation.

17. Another example of key public figures lending their credibility publicly to domestic companies and industries in a potentially very costly manner took place in the United Kingdom in 1990. At the outset of the Mad Cow crisis in Great Britain, Minister of Agriculture, John Gummer, had his young daughter Cordelia eat a beef burger in front of journalists on TV. He did this in an effort to convince the people of the country, as well as internationally, that there was no danger in consuming British beef. As British beef was later subject to extensive food bans, Gummer's actions were largely received as a symbolic ploy by a disaffected public (Grönvall, 1998:280).

18. Uriel Rosenthal and Paul 't Hart, prominent researchers of crises at Leiden University, go so far as to say that it is "no exagerration to say that every part and parcel of crisis management situations evokes expert advice" (Rosenthal and 't Hart, 1991:352).

19. The program of economic reform and dissolution of the Ministry of Works and Development into the department of WORKS, followed by its sell-off into splinter state-owned corporations is described in greater detail in the section on electricity reform.

20. What was perceived as the close association of local political actors with Mercury Energy is discussed from a symbolic and leadership perspective in the sections, *Leadership and Group Dynamics* and *Symbolic Politics Versus Technical-Managerial Logic*.

21. Drawing on the work of Kasarda and Parnell (1993), Jacobs notes that in less-affluent nations, "problems are commonly associated with rapid growth; social inequalities; economic restructuring; volatile global flows of corporate investment; demographic change; and environmental neglect" (Jacobs, 1997:127). While a distinction between types of urban crises on these grounds is certainly useful, defining characteristics should be flexibly conceptualized as features of urban crises in "developing" nations can be found in "developed" nations and features of "developed" in "developing" as well.

22. The literature on urban crises also differentiates between "generalized" and "sectoral" crises. While Jacobs notes that "[g]eneralized situations often encompass sectoral emergencies that are, themselves, symptomatic of underlying urban social, economic and environmental conditions" (Jacobs, 1997:128), sectoral crises represent "worst-case scenarios" where a number of serious urban problems converge and have a broad impact (ibid.)

23. "Soil thermal resistivity is a measure of the resistance to heat flow through a unit area of soil per unit of time, for a unit of thermal gradient, or ∞C m/W" (Ministry of Commerce, New Zealand, 1998:65). The measure "relates to the ability of the environment surrounding the cables to dissipate heat from the cables" (Mercury Energy Limited, Factual Report on Cables, 1998 and Integral Energy Cable Failure Report, 1998, cited in Ministry of Commerce, New Zealand, 1998:50).

24. Additionally, two reports produced outside of New Zealand are discussed briefly as well as examples of vicarious learning in crises: (1) Deputy Director of Victoria State Emergency Services (VICSES) Gareth Davis' Australian report *Visit to Auckland to Observe Power Crisis* and (2) a Swedish report produced by Lennart

Göhl of the State Energy Authority (*Energimyndigheten*) entitled *New Zealand's Electricity Market: Regulations Scrutinised Following the Serious Power Outage in Auckland, 1998* (*Nya Zeelands elmarknad: Regelverket revideras efter det allvarliga elavbrottet i Auckland 1998*). Terms of Reference or objectives of the inquiries are stated as well as the recommendations or conclusions at which the reports arrived.

5

Lessons for Crisis Managers

We will conclude this study by briefly identifying some of the most important lessons that may be drawn from the case experience. Our findings from the Auckland Power Crisis will be placed in a broader perspective through comparison with other cases documented in the international literature on crisis management.

5.1 EXPECT THE UNEXPECTED

Charles F. Hermann (1963), one of the founding fathers of modern crisis research, proposed surprise as a defining and stress-inducing condition of crisis management. In this case as in all too many others, "impossible" or unthinkable events occurred, overturning the conventional wisdom regarding the issue area in question (Rosenthal, et al., 1989:4–5; Sundelius, Stern, and Bynander, 1997; Stern, 1999:204–5). For example, the 1986 Chernobyl Fallout Crisis flew in the face of post-Harrisburg planning assumptions that nuclear accidents were likely to have local rather than transcontinental consequences. Major technical failures and disasters often occur at the horizon of scientific knowledge and may well, as in this case, be the result of complex systemic effects and interdependencies poorly understood by those in charge of large scale sociotechnical systems (Perrow, 1999; Sagan, 1993 Jervis, 1997:6). In Auckland, the system was designed to be robust enough to handle failure of one or two key cables (N-1 or N-2) without leading to major disruption of supply. However, the contingency that occurred was a cascading failure of no less than four of the main cables feeding the CBD, an unexpected N-4 contingency. As a result, there was little or no planning for the contingency which occurred and much of the public and private response was developed on an ad hoc basis.

The historical record suggests that, difficult as it may be, it is necessary for crisis planners and crisis managers to think the unthinkable and take even seemingly low probability (but high negative impact) contingencies seriously, at least as a basis for crisis prevention, preparedness, and planning activities. In other words, it is necessary for practitioners to get past the comforting idea that "it won't happen here" and confront the unsettling questions of "What if it does happen here?" and "Will we be ready?"

5.2 FACE THE WORST CASE SCENARIO

Even when the unlikely or the "impossible" has occurred, it remains tempting for embattled policymakers to continue to engage in denial or wishful thinking (Janis and Mann, 1977; Vandenbroucke, 1993) and to operate on the basis of a rosy scenario. While overreactions can be liable to criticism as well, experience suggests that elite and mass publics tend to be far more forgiving of overreactions than under-reaction.[1] It is much easier to be in the position of saying that precautions taken proved superfluous ("it wasn't as bad as we thought") than to say that opportunities to mitigate an emerging contingency were missed ("it was worse than we thought"). In crisis preparedness, erring on the side of caution is not a bad idea, even if this entails a degree of "inefficiency" (Landau and Chisholm, 1995).

The case of the "millennium bug" (Y2K) scare is an interesting point of departure in thinking about the implications of alternative scenario postures and particularly tolerance of vigilant overreaction. Many governments (and private corporations) did in fact act on worst case scenarios and make considerable investments in attempting to mitigate information system disruptions and secondary accidents in conjunction with the turn of the millennium. As it turned out, the apocalyptic scenarios discussed did not materialize. In general, it seems that the steps taken by governments were seen as responsible measures by mass publics and elite commentators alike and it does not appear that the proactive governments were punished for their vigilance in preparing for this contingency.

The experience from the Auckland power crisis speaks to this question as well. In fact there was a strong contrast between Mercury Energy—the stricken power provider—and the Police and Fire Services with regard to use of best/worst case scenarios. Mercury Energy operated on a rosy scenario for most of the crisis and was consistently forced to negatively revise its prognoses. The Police and Fire Services operated from worst case scenarios and acted proactively to deal with potential problems, a number of which did not materialize in the crisis. While it is clear that the positions of these different actors in the Auckland drama are rather different, the emergency services who overreacted exited the crisis with enhanced public confidence while Mercury Energy (and Mayor Les Mills who based his own reassuring public statements to a large degree on Mercury Energy's overoptimistic prognoses) left the crisis with serious credibility losses.[2]

5.3 TAKE CARE OF YOURSELF AND YOUR PEOPLE

A crisis, particularly a prolonged one such as the Auckland power crisis, is a physically and psychologically demanding ordeal for crisis managers and "foot soldiers" alike. The levels of physical and psychological stress experienced may be high if not extreme. In such situations, it may be very dangerous to assume that individuals are fully conscious of their own limits.

Temptations to push staff to the limit based on prognoses that the crisis is likely to be of short duration can be counterproductive if the crisis persists substantially longer than initial assessments predicted. The Auckland power crisis was by means unique in this respect. The tendency to "go all out" too soon is well-documented in the crisis literature (Stern, 1999; Steen et al., 1987; George, 1986).

Furthermore, burnout of leaders or key (and in some cases irreplaceable) staff can lead to serious performance decrements as a crisis wears on (Hermann, 1979; Vertzberger, 1990; Flin, 1996). Thus an important task of leadership is to face these limitations and husband human resources. Part of leading may be to exhort staff to perform beyond their normal efforts and capacity. But it is also to recognize when team members—including the leaders themselves—are approaching the point of exhaustion and see that they get the rest they need. The evidence from this case suggests that Mercury Energy demonstrated an exemplary sensitivity to this problem, monitoring the well-being of staff, providing them with access to rest and food onsite, and minimizing collateral stress by, for example, providing child care resources to those whose normal planning was disrupted by working overtime.

5.4 GOVERNMENTS CAN BE VICTIMS; VICTIMS CAN BE CRISIS COPERS

It is easy to overlook the fact that governments, aside from bearing responsibility for crisis response, are often victims of the crisis as well. Vital functions, lines of communication, or key players may be disrupted, obstructed, or otherwise put out of commission by the crisis. For example, during the Lima Hostage Crisis of 1996–1997 (Ullberg, 1998) several key government officials (such as the Foreign Minister of Peru) and relatives of the leadership (e.g. President Fujimori's own brother) were among the VIP hostages held by the Tupac Amaru guerrillas in the four-month siege of the Japanese Ambassador's residence. Government officials and agencies may be severely disturbed by natural disasters as well. In the Sudavik avalanche in Iceland to take another example, the local civil defense organization was incapacitated and had to be replaced by counterparts from a neighboring municipality (Bernardsdottir, 1998).

Similar problems beset the crisis managers in the (cf. Turner, 1978; Turner and Pidgeon, 1997) Auckland power crisis. The case demonstrated the dependence of modern government agencies and services (and that of private corporations) upon electricity for lighting, climate control, sanitation, transportation, communications, and data access/processing. Many facilities, especially those in high rise office towers, proved unusable and officials were forced to relocate. The bulk of the City Council staff, and the mayor himself, ended up relocating in order to get access to reliable infrastructure and decent working conditions. As we have seen, even secondary functions such as the courts had to be moved to improvised temporary facilities in order to continue operating. Yet these relocations were also time- and resource-consuming efforts, which added to the workload at a critical juncture in time. Some of these ad hoc moves, such as the mayor's use of Mercury Energy's premises for press conferences, created political—as opposed to operative—difficulties.

It is also important to keep in mind that not only can government be a "victim" of crisis but also that "victims" can transform themselves into effective and constructive crisis actors. For example, during the 1997 flooding of the Oder River in Europe, many reports suggest both more and less organized volunteer efforts (ranging from spontaneous efforts by individuals to groups such as the Scouts and the Red Cross) contributed significantly to mitigating the consequences of the flooding and helping with the relief effort (Chmielewski, 1999). Other recent historical examples include the Canadian Ice Storm Power Crisis and the Red River Flooding of 1998 (Newlove, 1998; Svedin, 1999). This was the case during the Auckland Power Crisis, too. Examples include a host of initiatives by Mercury Energy itself, by small business owners many of whom adapted to the crisis with creativity and a sense of humour, and large corporations who successfully relocated operations to the suburbs, to Wellington, or in some cases even abroad.

Thus, the self-sufficiency, resiliency and cooperative potential of individuals, groups, corporations, and non-governmental organisations and their potential contribution to crisis management should not be underestimated (Drabek, 1986).

5.5 TAKE "POLITICAL" RESPONSIBILITY SENSIBLY

Another challenge of crisis management concerns the art of taking political responsibility appropriately. This skill requires striking a delicate balance and steering clear of twin dangers. As has been emphasised in the literature on public relations, constituents (and consumers) tend to be very sceptical of crisis managers who deny any responsibility for negative events taking place on their watch or who clumsily attempt to shift all of the blame to other actors. For example, in incidents with at least great surface similarity—the Soviet downing of the South Korean (KAL) airliner in 1983 and the U.S. downing of the Iranian airliner in the Persian Gulf in 1988—world opinion was much more harsh in its condemnation of the USSR than of the United States. This was at least in part due to the contrast between the Soviet denial of responsibility for the KAL incident and the prompt and full assumption of responsibility by U.S. President Ronald Reagan for the tragedy in the Gulf (Cannon, 1991:318).

However, it is also essential for crisis practitioners to avoid being saddled with political/organizational responsibility for aspects obviously beyond their control or for decisions made well before they came to power. To take another example, the U.S. involvement in Vietnam was initiated by John F. Kennedy, and was subsequently inherited by successors Lyndon B. Johnson and Richard M. Nixon. By not adequately distancing themselves from the policies of their predecessors, the latter two presidents came to be blamed for the disastrous war (Gelb and Betts, 1979).[3]

In the Auckland case, Mayor Les Mills—by his use of Mercury Energy's facilities, his expressions of confidence in the power company's ability to fix the problem, and his policy of not criticizing Mercury—became very closely linked with the power company in the public mind. As a result, Mills was tainted by the negative image generated by Mercury's persistent inability to restore power during the crisis period. Our analysis suggests that the mayor came to assume more political responsibility for the operational performance of the power provider than necessary. A somewhat more agnostic and critical posture would have served his political career (and his constituents) better.

The Auckland experience is suggestive. Decision makers must be wary of taking "political" responsibility, albeit only implicitly or by association, for the outcomes of technical or political processes that are clearly beyond their sphere of competence and ability to influence. Failure to observe caution in this regard may prove very costly in terms of credibility.

5.6 DON'T LOSE SIGHT OF THE LONG-TERM PICTURE

In crisis decision making and crisis communication it is essential to balance long-term and short-term interests and priorities. A convincing body of laboratory and field results testifies to the tendency for decision makers to focus on the short-term picture in crisis situations. This is a common psychological reaction to heavy stress and one that can have both functional and dysfunctional consequences depending upon the circumstances (Hermann, 1979; Holsti, 1989; Flin, 1996). Yet, awareness of this tendency makes it possible to take steps to combat the negative effects and strike a better balance between the acute needs of the situation and the longer-term values and interests at stake. One tactic which may be helpful is to delegate responsibility for longer term strategic thinking to one or more officials who are protected from the overwhelming burden of problems which need to be solved immediately and allowed the "luxury" (actually more of a political and policy necessity) of contemplating the relationship between the immediate and the long run.

In fact, a number of actors in the Auckland case did seem to be able to maintain this kind of focus. For example, our findings show that both the Ministry of Civil Defence and the City Council's civil defense staff were aware that the management of this crisis could have important implications for existing controversial plans to reform the Civil Defence Act and civil defense system. In other words, the practitioners maintained conscious of the fact that this experience was both a "test case" and a potential precedent. This is reminiscent of the thinking in the case of the stranded Soviet submarine, where Swedish decision makers were very conscious of the fact that Swedish behavior in the acute crisis might have dramatic implications for the viability of then highly-valued neutrality policy (Theutenberg, 1982; 1986; Stern, 1990). In contrast, it has been suggested that during the Swedish response to the Chernobyl disaster (Stern, 1999), inadequate attention was paid to longer-term strategic issues during the acute phase of the crisis (interview with former Swedish Radiological Protection Institute Director General Gunnar Bengtsson, 12/12/96).

5.7 BALANCE SYMBOLIC AND OPERATIONAL CONCERNS

't Hart (1993) emphasises the importance of distinguishing between symbolic-political (SP) and technical-managerial (TM) logics when designing and assessing crisis management. It is quite common for these very different desiderata to diverge in crisis situations—measures or utterances which may be appropriate from the SP perspective because they are appealing to mass or elite publics, easy to explain, or may be useful in solving political problems of one kind or another, may pose serious difficulties from a narrower operational perspective—and vice versa. For example, during the Swedish Chernobyl crisis, Prime Minister Ingvar Carlsson chose to remain on the island of Gotland during the first days of the crisis. He reportedly felt that the crisis was being effectively managed and that, not possessing any particularly relevant technical or medical knowledge, that he had nothing to add. From the TM perspective, this choice makes perfect sense (Stern, 1999). However, from the SP perspective, an important part of Prime Ministerial leadership might well have been to symbolically indicate that the issue was being taken seriously at the pinnacle of government and to comfort a shocked and fearful nation.

Similarly, Premier of the province of Quebec Lucien Bouchard demonstrated mastery of the SP side of crisis management during the 1998 Ice Storm Power Crisis in Eastern Canada. Bouchard maneuvered Canada's federalist Prime Minister Jean Chrétien into an awkward position by leading his rival to believe that both figures would attend a trade mission to Latin America during the crisis. Chrétien exposed himself politically by leaving the country during an emergency. Bouchard then cancelled his trip; a gesture of political responsibility and solidarity made all the more vivid by the Prime Minister's absence (Newlove, 1999).

The Auckland case contains several good examples of actions or proposals that demonstrate the potential tension between the TM and SP dimensions of crisis management. The mayor's decision to use Mercury's facilities, to speak from their podium, framed by the company logo, made perfect sense from the TM perspective. It solved his urgent infrastructural problems and kept him close to the available technical expertise. From a SP perspective, the very same move was a disaster. It became more difficult for Mills to criticize Mercury, diminished his stature and aura of independence, and made him seem a bit too close to Mercury for the comfort of many observers—including his own public relations advisor (May, interview, 11/18/98). Furthermore, Mills became symbolically tarnished by this association with the "villain" in this public drama. By contrast, the City Council staff's acquisition of a portable freezer container in a vain attempt to save perishable foodstuffs made good symbolic-political sense. However, the portable freezer container was a disaster waiting to happen from an operational food protection perspective—one that could easily have resulted in serious health risks had the public health experts (who were very much opposed to the idea on TM grounds) not forcefully intervened.

Crisis participants need to ask themselves not only whether a given action or statement will work but also how it will look and be received in order to more skilfully navigate the stormy seas of crisis management.

5.8 BEWARE OF CREDIBILITY TRAPS

Several of the previous points have touched upon the concepts of credibility and credibility traps. A credibility trap is a pitfall in which a crisis actor gambles with his or her credibility through myopic framing or by making ill-advised commitments to other actors, stakeholders, or the public at large. So-called spin (Smith, 1988; Matthews, 1988) is a crucial element in avoiding credibility traps. Actors can often influence the ways in which their actions and utterances are judged via suitable framing and benchmarking of crisis behavior. For example, actors can emphasize that the problems they are dealing with are very difficult, and thus lower expectations somewhat. Conversely, suggesting that the problems are minor (and preparedness good) tends to increase public and media expectations.

There are a number of variants of credibility traps. A classic form stems from the rosy (as opposed to worst-case) scenario posture mentioned above. When actors present constituents with optimistic prognoses, their future credibility becomes a hostage to the accuracy of the prognosis in question. Mercury Energy and, to a lesser extent, Auckland Mayor Les Mills fell prey to this trap as did the Swedish National Radiological Protection Agency (which initially and mistakenly advised citizens that intrusive interventions would not be required) in the Chernobyl Fallout case (Stern, 1999). The police and fire authorities in the Auckland case, as we have seen, did not fall into this trap, but rather successfully protected their credibility via a proactive and vigorous response based on worst-case scenarios.

Another credibility trap has to do with the temptation to be decisive. Under pressure from the media and or other crisis actors, it is tempting for leaders to make strong statements or commitments about crisis or post-crisis policy. Yet crisis issues tend to be characterized by value complexity, technical and or political complexity, and persistent uncertainties of various kinds. As a result, it may be very difficult to make good or even sustainable decisions "shooting from the hip." Promises to restore power quickly (Auckland), to salvage a sunken ferry and recover the bodies (the *MV Estonia* disaster in the Baltic Sea), to capture a submarine intruder in Swedish territorial waters (Hårsfjärden), to complete a controversial tunnel (Hallandsås), or to rid Sweden of all outlaw motorcycle gangs (the Nordic MC Wars) may prove embarrassing and painful to actors and stakeholders alike if, as in these cases, it turns out to be impossible to deliver in the short- to medium-term.[4] Richard Neustadt (1990) has argued that leaders must weigh the effect of each individual action on their future political effectiveness if they are to remain politically potent in the long run. Similarly, each statement or commitment made by crisis managers should be considered in terms of its long-term effect or potential effect on credibility. Credibility is a sociopolitical resource that is difficult and time-consuming to amass, but may be squandered in an instant, and is difficult and sometimes even impossible to rebuild.

5.9 MAKE SURE YOUR MESSAGE IS GETTING OUT

Crisis communication is a particularly demanding task. Many organizations are ill-prepared for this pressure psychologically, in terms of human resources and with regard to vital infrastructure adequate to the heavy demands of crisis operations. Normal information units generally cannot cope with the pressure and must be reinforced from other parts of the organization or with outside help. In some cases, the organization has planned and practiced for this reinforcement, in other cases not. Either way, the pace and pressure are often greater than anticipated and it may be a constant struggle to try to keep up, and in the best case, take the initiative in crisis communication. Under these conditions, crisis communicators can all too easily fall into a reactive mode of "fighting fires" as they flare up, losing track of the evolving big picture.

Planned (in the best case) activities such as systematically monitoring news broadcasts and newspapers in order to see how the "message" is being received may seem like a dispensable luxury. Yet this seemingly less urgent task is a crucial one. Only by monitoring the media (and attending to direct communications from various publics) can the organization maintain sensitivity to feedback. This kind of feedback is essential if the organization is to quickly identify and correct information distortions, unsubstantiated rumours, and respond to developing public concerns. The available evidence suggests that Mercury Energy did succeed in producing an impressive body of output (press releases, public information advertisements, press conferences, etc). Unfortunately, some of this information contained contradictions (described in the sections above on the cable failures) which disturbed consumers. Furthermore, the organization had difficulty in sustaining its media monitoring function, which in turn slowed down the organizational learning process and cost the organization credibility over the course of the crisis. The same holds true for the mayor who did not detect his mounting credibility problems in time to prevent them from becoming very serious indeed. Once again, this experience coincides with the Swedish National Radiological Protection Institute's experience during the Chernobyl nuclear accident where planned media monitoring functions were de-prioritized in favour of other concerns, leaving the organization at least partially "blind" to its growing credibility problems (Steen et al., 1987; Nohrstedt, 1991; Stern, 1999).

5.10 TRY TO WORK WITH THE MEDIA

As we have noted above in the section on crisis communication and media management, the massive power of the media is a fact of life for crisis managers. Regardless of the facts regarding crisis performance, the portrayal in the media is often decisive in terms of the political repercussions of a particular episode. Even the most effective crisis management effort can come to be seen as a political-administrative failure if portrayed as such in a sustained fashion by the media. Crisis and media practitioners have both convergent and divergent interests in and after crisis situations. Decision makers and their information staffs have access to information, which they need to disseminate. Journalists need this information too and as professionals, generally wish to pass it on. On the other hand, journalists also have a duty to assess critically the crisis response, which sometimes places them at odds with the crisis practitioners. Furthermore, from a commercial perspective, journalists (even serious journalists) are drawn to the dramatic elements, which make for good stories. Sometimes this orientation leads to neglect of more mundane (yet important) information that crisis managers want to communicate to the public via the media. It can also lead to an emphasis on the "spicy" aspects of the case to the neglect of other aspects.

Many handbooks of public relations emphasise cultivating good personal relationships between information or top-line staff and the media. Such relationships are of course, likely to be helpful when a crisis occurs. However, the sheer volume of media attention in crises is such that it is likely that the bulk of media contacts will be with relatively unfamiliar journalists. This is partly due to the fact that those regularly covering a particular "beat" will most likely be reinforced by generalists in a crisis situation. Similarly, if normal media relations tend to be primarily with local or locally-based representatives of national media, in a crisis situation it is likely that a higher proportion of national and international media will turn up. Since one cannot count on prior relationships as adequate, attempting to build a constructive relationship with the journalistic corps during a crisis is extremely important.

Getting off to a good start is very helpful. A cold- or slow-started initial reception of media inquiries, or one that is arrogantly disrespectful of naive early questions, can be very costly. John Long, information officer of the Oklahoma City Fire Department describes how he nearly made a costly blunder during the crisis following the bombing of the Alfred P. Murrah Federal Building in 1995. During a hectic press conference in the midst of the rescue operations, Long was asked by an unknown reporter whether it was true that the rescue dogs were suffering from depression. Long—inwardly scornful of what appeared to be a silly question in a tragic situation—stifled the temptation to make light of the query and answered that he had no knowledge of canine depression, but would look into the matter. It turned out that the unknown reporter was covering the story for the influential *New York Times*. Furthermore, it turned out that the rescue dogs—used to rescuing living "victims" in drills—were disheartened by the heavy death toll. Play sessions with the dog handlers were inserted into the mission schedule in order to maintain the dogs' spirits and improve their performance. Long narrowly missed getting off to a bad start indeed.[5]

In the Auckland case, we found that the fire department—benefiting from good prior public and press relations developed during an ongoing public safety campaign—got off to a good start and maintained friendly relations with the media throughout the crisis. The police, who had a more troubled relationship with the media prior to the crisis, also managed to get off to a good start and improve their relationship with the media and the public at large during the crisis. By contrast, Mercury Energy got off to a relatively bad start, alienating elements of the press corps and contributing to sustaining a fault-finding mentality among the many of the top journalists. Given the character of the crisis development process (in which a sequence of events that Mercury had assured the press and the community was virtually impossible actually occurred), Mercury started with something of credibility deficit, which worsened as the crisis wore on. By the time Mercury's information effort came fully online (and it was exemplary in some respects), the invaluable benefit of the doubt had already been lost and many journalists were already on the proverbial warpath.

As we have indicated, Mercury's information effort became increasingly professional as the crisis wore on. For example, in response to journalists initially placing cameras on poles in order to attempt to get in interior shots of the Mercury Premises, a press area inside the building was subsequently made available. This was a good adaptation to the adage that one should give the press what they need (if not, they will take it anyway). Of course, voluntary provision of access tends to garner some good will with media practitioners, where as covert or forced access does not. This is also the case with breaking bad news or disclosing ones own shortcomings, as noted above in the sections on taking responsibility sensibly. In fact, after the initial period, considerable efforts were made to make communications infrastructure and up to date information available to journalists, which is in keeping with good public relations practice.

5.11 A WORD OF CAUTION

The reality of multidimensional crisis management in complex political and organisational environments is such that such sweeping generalizations are likely to be misleading. The players in the drama of the crisis we have studied have differed substantially from each other in their performances. In fact across the crisis, the very same actor may have performed well with regard to some issues and less well with regard to others. So we will avoid the temptation to end this study with an aggregate assessment. At the end of the day, the actors we have studied were forced to fumble around in the dark, dealing with thorny and often technically complex issues, trying to muddle through without the benefit of the hindsight which informs (and occasionally distorts) the assessments of researchers.

In this final chapter, we have looked for some potential dos and don'ts. We have tried to place our findings from the Auckland Power Crisis in their proper perspective through a series of comparisons with other cases from the literature. We do not suggest that the observations and rules of thumb presented in this chapter should be seen as a simple recipe for a successful crisis experience. Rather we hope that the experience we have described and analyzed will help to sensitize potential crisis managers to some important things that can go wrong and right in coping with crisis.

NOTES

1. See Chmielewski's (1999) analysis of the Polish response to the Oder flooding of 1997 and Stern (1999) on the Swedish management of the Chernobyl Fallout Crisis of 1986 for examples of how rosy scenarios can impede and delay a crisis response.

2. It should be noted that Mayor Les Mills, following the advice of the Emergency Services experts, did act on the basis of a worst-case scenario when he initially asked the people of Auckland to stay out of the CBD. While this move was later criticized, this criticism focused on the fact that the Mayor negelected to specify for how long citizens and merchants should avoid the CBD.

3. By contrast, Swedish Supreme Commander Owe Wiktorin acted decisively in the case of the submarine-mink confusion (Brändström, 2000) proactively disclosing

that the Swedish defense had confused sounds made by swimming minks with foreign submarine activity. Wiktorin dealt with this embarrassing misperception early in his tenure (while it was still obvious that the bulk of top-level responsibility for this issue belonged not to him, but to his predecessor). In so doing, he avoided being tagged with personal responsibility for this performance failure. In addition, by coming forward voluntarily with the information (rather than waiting for the media to get wind of the problems), Wiktorin was able to keep the initiative and minimize the credibility damage to his organization.

4. See SOU 1996:189; Bynander, 1998; Kärde, 1999; Svedin, 1998.

5. A presentation made at the conference "Crisis Management at the National Level" in Skåvsjöholm, Sweden (Long, 1996).

Appendix: Chronology of Events

Tuesday, January 20

Control cables (associated with 110kV No.1 feeder from Transpower's (TP) Roskill Substation to Mercury Energy's (MEL Liverpool Substation) were inadvertently cut, by a contractor working at Liverpool Station, which caused the automatic disconnection of the No.1 110 kV circuit. The parallel No.2 circuit was thus overloaded, resulting in its automatic disconnection, and total loss of supply to Victoria, Newton and Liverpool Zone Substations, normally supplied by the Liverpool 110/22kV Substation. While full restoration of supplies was completed within about forty minutes through re-arrangement of the electrical network, during this process the loading of gas cables on the Penrose-Quay 110 kV cables rose to the maximum specified rating (normal operating levels were about 50% of the specified rating). According to the Ministry of Civil Defence Report, the load on the gas filled cables rose for a short time to about 53MVA, which was within their continuous rating of 60MVA.

Thursday, January 22 (No. 2 Gas Cable Failed)

A fault in the Penrose-Quay 2 110kV Impregnated Gas Pressure Cable (IP) occured (at 12:29 P.M. under a load of 30MVA) and while not resulting in loss of supply, caused the parallel Penrose-Quay 1 110kV IP cable to exceed its specified rating by 20% for a short time (it rose to 58MVA and then was reduced to 41MVA).

Monday, February 9 (No. 1 Gas Cable Failed)

Penrose-Quay 1 110 kV IP cable developed a full fault at 6:01 P.M. at a load of 42MVA. (Only customers supplied by Hobson Zone Substation were cut off for around thirty minutes due to previous system rearrangements). CBD customers were asked to reduce consumption by 10% and stand-by generation as well as reconfiguration and strengthening of the system were introduced in the (assessed as weakest) Quay area. This fault and subsequent actions to restore power led to the overloading of the Roskill-Liverpool 110kV oil cables by 30% over normal operating limits (20% below maximum specified limits).

Friday, February 13

Mercury announced the construction of a temporary 22kV overhead line from Liverpool to Quay Substations under emergency regulations, in order to protect against the failure of the existing tie line between the two Substations.

Tuesday, February 17

A contractor accidentally severed some controller cables at the Liverpool Station, which were attached to the Mt. Roskill station (one of the points of entry for power that were to be distributed through Mercury's transmission nets in Auckland).

Thursday, February 18

Mercury Energy put out two daily press releases in which the company, among other things, conveyed the tenuousness of the system and asked for reductions in the use of power. These appeals continued the following day.

Thursday, February 19 (No. 1 Oil Cable Failed)

Fault occured on the Roskill-Liverpool 1 110 kV oil cable at a load of 51MVA (which had a rating of 60MVA under normal, i.e., both cables working circumstances), automatically disconnecting both cables and resulting in total loss of supply to Liverpool, Quay, Newton and Hobson Zone Substations. The effects were so severe due to the network still being in a reconfigured state following the gas cable failures originating at Penrose-Quay. Only one cable faulted, but the protection systems covering it didn't work and thus both cables were automatically disconnected. Load on the remaining in-service cables increased by around twelve percent above normal. According to the Ministry of

Civil Defence Report, the remaining oil filled cable rose to 70MVA, which should have not constituted a problem as it was within its short-term rating. Mercury confirmed that 70MVA was the continuous rating for this remaining cable as a single line.

Friday, February 20 (No. 2 Oil Cable Failed)

Despite introduction of a customer load management program introduced between 8 A.M. to 5 P.M. to keep the remaining in-service oil cables on the Roskill-Liverpool line stabilised at around twelve percent above normal, at 6 P.M. a fault in the remaining Roskill-Liverpool 110kV 2 developed at a load of 61 MVA and all supply to Liverpool, Quay, Newton and Hobson Substations was cut off. The network was again reconfigured, with as much load as possible transferred to zone substations further out from the CBD, and the 22kV backup from Kingsland to Liverpool established. Mercury asked the Police Dept. about declaring a state of emergency, not realising it is a matter for the city civil defence organisation. Apparently, police first thought it was a joke. At 8:30 P.M. the power-outage caused fire alarms to go off all over the Auckland Central Business District, bringing the problem to the Fire Service's attention. A first emergency meeting was held at 11 P.M. at Mercury Energy's office together with the emergency services and representatives of the Auckland City Council. Mercury Energy started searching nationally and later internationally for back-up generators.

Saturday, February 21

The first comprehensive response meeting was held at 10 A.M. at Mercury's premises with all stakeholders invited. Mercury Energy and the Mayor asked the Fire Service if they could cope, or if they thought a state of emergency should be called; Emergency Services argued that they could cope with the situation. The Auckland Healthcare Representative present thought the issue should have been more thoroughly discussed. The Auckland City Council spent the remainder of the two days following the failure of the fourth cable relocating to alternative office spaces. Vital functions situated in the CBD, such as the District and High Court of Auckland, spent the weekend finding alternative locations.

Sunday, February 22

The Mayor of Auckland asked people, in a public statement at 12:15 P.M., to stay out of the CBD due to the power cuts in the area. Another statement was issued at 5:15 P.M. arguing that "Auckland's CBD power situation should not become a Civil Defence Emergency as people appear to have followed the advice of emergency services".

Sunday, March 1

Mercury Energy and the contractor they have hired started surveying the railroad lines leading into the CBD for the building of a new overhead power line that could feed power into the district.

Tuesday, March 3

Attempts to repair the Roskill-Liverpool 2 oil cable following the February 20 fault proved unsuccessful. Tests of the cable on March 3 resulted in a failure at approximately 50 kV.

Wednesday, March 4

Attempts to repair Roskill-Liverpool 1 oil cable following the failure on February 19 also proved unsuccessful. Tests of the recommissioned cable resulted in failure at about 80 kV.

Friday, March 6

An article in the Auckland Herald stated that the area affected by the outage included the workplaces of around 63,000 individuals in 10,000 companies, as well as about 4,000 residents.

Monday, March 9

Cabinet agreed to a Ministerial Inquiry by the Ministry of Commerce looking into the technical aspects of the crisis (CAB (98) M 8/4).
The Ministerial Inquiry is mandated with two basic tasks:

1. What caused the power supply, and in particular, the power cables into the CBD of Auckland, to fail?
2. In light of the cable failure, to recommend changes to ensure that security risks related to Mercury Energy's distribution lines are managed efficiently.

Mercury Energy rented the ship Union Rotorua docked in the Auckland harbour to create extra supply of power for container cranes.

Monday, March 11

The Director of Civil Defence, at the request of the Ministers of Civil Defence and Energy reviewed the merits of a declaration, concluding that a declaration was unnecessary, but added his reservation that "My greatest concern is, and always has been, that no one is seen as being in charge at the local level".

Thursday, March 19

Mercury released a statement via e-mail stating that there would be "9000 residential and business customers in the affected parts of the CBD in the coming week".

Monday, March 23

The Minister of Civil Defence, the Hon. Jack Elder, wrote to the Prime Minister and explained that a state of emergency declaration was unnecessary as one of the conditions (that local agencies cannot cope with the situation) as it is specified in the Civil Defence Act 1983 had not been fulfilled.

Wednesday, March 25

The Secretary of Civil Defence, at the request of the Director, approved the terms of reference for a separate (from the Ministry of Commerce report) but concurrent report reviewing the crisis. The aim of the Ministry of Civil Defence's report was to improve emergency management. The main questions posed were as follows:

1. Was there a need to declare a state of civil defence emergency?
2. Was there an effective response to the crisis by the key responders?
3. How adequate was the public information process?
4. What was the contingent capability to overcome power disruption?
5. Are there other critical supply or service provision issues which were, or could have been, affected by the crisis and which may be applicable to other local authorities?
6. What was the contribution made by the Ministry of Civil Defence staff? Did it add value?

As a result of a meeting between the Director of Civil Defence and the Deputy Secretary for Commerce, the fourth question was rephrased in order to eliminate overlap, as follows: What was learned about the availability of portable generation to substitute for grid electricity?

Second Week of April

The construction of the overhead power line into the CBD was finished and put into use terminating the immediate problem with the lack of power in the area.

Monday, April 27

The Mayor publicly announced that businesses and residents could start moving back to the CBD. Plans were made to ensure chaos did not arise as people rushed back but the problem turned out to be that virtually no one showed up. Conscious efforts to attract people and businesses back to the CBD were intensified.

Friday, May 8

A technical report prepared by Integral Energy of Australia regarding the first question of the Ministerial Inquiry was presented to the Minister of Energy, Hon. M. Bradford.

Sunday, May 10

At 5:41 P.M. the Roskill-Liverpool 2 110kV oil cable failed, leading to a fire in the joint box.

End of June

The trust in charge of the funds allocated to get businesses and people back to the CBD completed its work and the board was dissolved.

Bibliography

Abelson, R., and Levi, A. (1985) "Decision Making and Decision Theory." In *The Handbook of Social Psychology*. Ed. G. Lindzey and E. Aronson. 3rd ed., 231–309. New York: Random House.

Allison, G. (1971) *Essence of Decision: Explaining the Cuban Missile Crisis.* Boston: Little, Brown.

Allison, G. and Zelikow, P. (1999) *Essence of Decision: Explaining the Cuban Missile Crisis.* 2nd ed. New York: Longman.

Anders, G. (1990) *Probability Concepts in Electric Power Systems*. New York: John Wiley & Sons.

Anderson, P. (1987) "What Do Foreign Policy Makers Do When They Make A Foreign Policy Decision? The Implications for the Comparative Study of Foreign Policy." In *New Directions in the Study of Foreign Policy.* Ed. Charles Hermann, Charles Kegley, and James Rosenau, p. 285–308. Boston: Allen and Unwin.

Armstrong, A. (1993) "Urban Crisis and Basic Services: Solid Waste Management in Dar es Salaam." *Ekistics. Oikistike* 60, nos. 358/359: 74.

Auckland City Council (1998a) The 1997/1998 Annual Report. Auckland, New Zealand.

Auckland City Council (1998b) The 1998/1999 Annual Plan: Long Term Financial Strategy Funding Policy. Auckland, New Zealand.

Auckland City Council (1998c) "Out of the Black and Into the Light" City Scene: The Voice of Outstanding Auckland, (30/03/98). Council News available at: http://www.akcity.govt.nz/news/cityscene/199803/300398/il.htm

Auckland Regional Council (1998) Caring For Our Region, Our Future: Strategic Plan of the Auckland Regional Council Auckland, New Zealand.

Australian Broadcasting Corporation, Online Edition of ABC News (1998) "Power Crisis Eases in Old" (02/24/98). Available at: http://www.abc.net.au/news/98/02/24/980224_13.html

Ayeni, V. O. (1991) "Managing Nigeria's Urban Crisis—an Overview of Current Strategies, Problems and Lessons," *Nagarlok* 23, no. 2: 1.

BBC News: World-Asia Pacific (1998) "Auckland's Heart of Darkness" (02/24/98, Published at 11:44 GMT). Available at:http://news1.thls.bbc.co.uk/hi/english/world/asia-pacific/newsid_58000/58781.stm

Bass, Bernard M. (1998) *Transformational Leadership: Industrial, Military, and Educational Impact*. Mahwah, N.J.: Lawrence Erlbaum

Beck, U. (1998) "Politics of Risk Society." *In The Politics of Risk Society*, Ed. Jane Franklin. Cambridge, U.K.: Polity Press.

Bernardsdottir, A. (1998) "An Avalanche in Sudavik in the Western Fjords of Iceland." Unpublished report. Crisis Management Europe: National Crisis Management from an International Perspective.

Billings, R., and C. F. Hermann (1998) "Problem Identification in Sequential Policy Decision Making: The Re-representation of Problems." In *Problem Representation in Foreign Policy Decision Making*. Ed. D. Sylvan and J. Voss. Cambridge, U.K.: Cambridge University Press

Birch, B. "Parliamentary Session of the Government of New Zealand" Hansard (02/24/98). Available at: http//rangi.knowledge-basket.co.nz/hansard/han/text/1998/02/24-019.html

Bishop, R. (1998) "Freeing Ourselves From Neo-Colonial Domination Research: A Maori Approach to Creating Knowledge," Qualitative Studies in Education, 11(2):199–219.

Blaise and Masicotte. (1996) In "Comparing Democracies: Elections and Voting in Global Perspective." Ed. Lawrence LeDuc, Richard Niemi, and Pippa Norris. Thousand Oaks, Calif.: Sage.

Blumler, J. and Gurevitch, M. (1995) *The Crisis of Public Communication*. London: Routledge.

Bolay, J.-C. and Du, T. T. N. (1999) "Sustainable Development, Urbanization, and Environmental Risks: The Priority of Local Actions in Ho Chi Minh City, Vietnam." *Journal of Urban Technology*, 1 August, vol. 6, no. 2: 65–85.

Boston, J. (1998) "Public Sector Management, Electoral Reform and the Future of the Contract State in New Zealand" Australian Journal of Public Administration vol. 57(4):32–44.

Boulle, P., Vrolijks, L. and Palm, E. (1997) "Vulnerability Reduction for Sustainable Urban Development," *Journal of Contingencies and Crisis Management* 5, no. 3: 179–88.

Bovens, M. and 't Hart, P. (1996) *Understanding Policy Fiascoes*. New Brunswick, N.J.: Transaction.

Bradford, M. (1998) "Parliamentary Session of the Government of New Zealand" Hansard (02/24/98). Available at: http//rangi.knowledge-basket.co.nz/hansard/han/text/1998/02/24-019.html

Brecher, M. (1974) *The Foreign Policy System of Israel.* New Haven, Conn.: Yale University Press.

Brecher, M. (1993) *Crises in World Politics: Theory and Reality*. New York: Pergamon.

Breslauer, G., and P. Tetlock, eds. (1991) *Learning in U.S. and Societ Foreign Policy*. Boulder, Colo.: Westview Press.

Britton, N. (1998) "Managing Community Risks." Wellington: Emergency Management Policy & Establishment Unit—New Zealand's Ministry of Civil Defence.

Britton, N. (1999) "Sustainable Hazard Management in New Zealand: Developing Practice Frameworks." Paper presented at the IDNDR Conference on Natural Disaster Prevention, Land-Use Planning, and Sustainable Development (June 17–19, 1999). Paris: Assemblee Nationale.

Britton, N., and G. Clark (1999) "Redesigning Emergency Management: Applying Principles and Practices in New Zealand." New Zealand: Ministry for Emergency Management.

Bruner, J. (1957) "On Perceptual Readiness." Psychological Review 64:123–52.

Brändström, A. (2000) *Minkar eller ubåtar i den svenska ubåtsjakten* [*Minks or Submarines—The Dilemma in the Swedish Hunt for Submarines*]. Crisis Management Europe—National Crisis Management from an International Perspective, Stockholm: Överstyrelsen för Civil Beredskap (ÖCB) [The Swedish Agency for Civil Emergency Planning].

Bush, G. (1997) "Local Government." In *New Zealand Politics in Transition*. Ed. Raymond Miller. New Zealand: Oxford University Press.

Businessweek (1992) "The Crisis of Urban America." No. 3266: 38.

Buzan, B. (1991) *People, States and Fear: An Agenda for International Security Studies in the Post–Cold War Era*. Boulder, Colo.: Lynne Rienner.

Buzan, B., O. Waever, and J. De Wilde (1998) *Security: A New Framework of Analysis*. Boulder, Colo.: Lynne Rienner.

Bynander, F. (1998) "Crisis Analogies. A Decision Making Analysis of the Swedish Hårsfjärden Submarine Incident of 1982" (Research Report 29). Stockholm: Swedish Institute of International Affairs.

Calvert, R. (1998) "Explaining Social Order: Internationalization, External Enforcement, or Equilibrium." In *Institutions and Social Order*. Ed. Soltan, Uslaner, and Haufler. 131–162. Ann Arbor: University of Michigan Press.

Cannon, L. (1991) *President Reagan: The Role of a Lifetime*. New York: Touchstone/Simon and Schuster.

Castenfors, K. "Swedish Older People's Rights to Social Care in War, Civilian Disaster and Societal Crisis." In *The Sociology of Aging: International Perspectives*. Ed. V. Minichiello, N. Chappell, H. Kendig, and A. Walker. Melbourne: International Sociological Association Research Committee on Aging, 1996.

Castenfors, K. and P. Stenström (1998) *Fredagen den 13e. En studie av FOA: s kris- och informationsberedskap med anledning av gasolyckan på Tegeluddsvägen i Stockholm i februari 1998*. FOA-R—98-00853-222—SE.

Castenfors, K. and L. Svedin (2000) *Samhällets förmåga att hantera svåra påfrestningar i fred* [*Society's Ability to Withstand Severe Pressures in Peacetime*]. Draft version. Stockholm: FOA Dnr 00-4486/S.

Chakrabarti, P. G. D. (2001) "Urban crisis in India: new initiatives for sustainable cities" *Development in Practice*, 1 May, vol. 11, no. 2: 260–72(13).

Chmielewski, P. (forthcoming) "The 1997 Flood Disaster in Poland." Crisis Management Europe—National Crisis Management from an International Perspective.

Chorianopoulos, I. (2002) "Urban Restructuring and Governance: North–South Differences in Europe and the EU URBAN Initiative" *Urban Studies*, 1 April, 39, no. 4: 705–26.

Coates, J. F. (2001) "Technology and the Changing City," *Journal of Urban Technology* 8, no. 2: 95–106.

Consumer Reports. (1998) "Electricity Deregulation: Consumer Beware." 63 No. 10:6.

Commerce Commission. (1999) "The Electricity Industry Reform Act 1998 Commission's Role and Processes" (Practice Note No. 3). (Original version, September 1998, revised version, August 1999). Available at: http://www.comcom.govt.nz/publications

Crump, S. (1966) *Black Riot in Los Angeles: The Story of the Watts Tragedy.* Los Angeles: Trans-Anglo.

Dabinett, G. (1999) "Technological Change and Urban Economic Development: Local Policy in Sheffield, United Kingdom (1981–1996)." *Journal of Urban Technology* 6, no. 2: 87–108.

Dalziel, P. and R. Lattimore (1996) *The New Zealand Macroeconomy: A Briefing on the Reforms*. Auckland: Oxford University Press.

David, C. (1993) *Foreign Policy in the White House*. Lanham, Md.: University Press of America.

Davies, D. (2002) "World Trade Center Lessons: Risk Management Lessons From the World Trade Center," *Computer Law and Security Report*, March, 18, no. 2: 117–19(3).

Davis, G. (1998a) "The Auckland Electricity Supply Distribution 1998: Emergency Management Aspects." Unpublished article intended for the *Journal on Auckland*.

Davis, G. (1998b) "Visit to Auckland to Observe Power Crisis." Preliminary report (unpublished) of Deputy Director Gareth Davis of Victoria State Emergency Service (VICSES), Australia.

Design Power Limited. (1999) "About Us: Design Power Parsons Brinckerhoff (PB)." Available: http://www.designpower.co.nz

DiPasquale, D., and E. L. Glaeser. (1998) "The Los Angeles Riot and the Economics of Urban Unrest." *Journal of Urban Economics* 42 no. 1: 52–78(27).

Donald, R. (1999) "National Plans for MMP Increase Political Instability." Green Party of Aeotearoa, New Zealand Press Release (April 25, 1999). Available: http://www.greens.org.nz

Douglass M. (2000) "Mega-urban Regions and World City Formation: Globalisation, the Economic Crisis and Urban Policy Issues in Pacific Asia." *Urban Studies* 37, no. 12: 2315–35.

Dror, Y., P. Lagadec, B. Porfiriev, and H. Quarantelli. (1999) "Crisis to Come: A Symposium" *From Crises to Contingencies: A Global Perspective*. Ed. Uriel Rosenthal, Louise Comfort, and Arjen Boin. Draft chapters, November 1999.

Dror, Y. (1988) "Decisionmaking under Disaster Conditions." In *Managing Disaster: Strategies and Policy Perspectives.* Ed. Louise Comfort. 255–273. Durham, N.C.: Duke University Press.

Dror, Y. (1986) *Policymaking under Adversity*. New Brunswick, N.J.: Transaction.

Durie, M. (1997) "Mana Maori Motuhake: The State of the Maori Nation." In *New Zealand Politics in Transition*. Ed. Raymond Miller. New Zealand: Oxford University Press.

Dynes, R. R. (1970) *Organized Behavior in Disaster.* Lexington, Mass: n.p.

Dynes, R. R., and K. J. Tierney. (1994) *Disasters, Collective Behavior, and Social Organization*. Newark: University of Delaware Press.

Eckstein, R. (1997) *Nuclear Power and Social Power.* Philadelphia: Temple University Press.

The Economist (1998) "The Coalition Crumbles." vol. 348, August 22–28.

Edelman, M. (1988) *Constructing the Political Spectacle.* Chicago: University of Chicago Press.

Edlin, B. (1998) "Absolutely Positively Privitization" *Management* vol. 45(5):96.

Elazar, D. (1994) *Federal Systems of the World: A Handbook of Federal, Confederal and Autonomy Arrangements,* (2nd ed.). Harlow: Longman Current Affairs.

Elder, J. "Parliamentary Session of the Government of New Zealand." Hansard (02/24/98). Available at: http://rangi.knowledge-basket.co.nz/hansard/han/text/1998/02/24-019.html

Elias, D. "The City That Went to Sleep" The Age, Melbourne Online (980228). Available at: http://www.theage.com.au

Emergency Management Group (EMG) Wellington. (1999) Planning Document of the Regional Emergency Management Group (EMG) in Wellington. Available by contacting Wellington City Council Emergency Management Office (EMO) http://www.wemo.wcc.govt.nz

Ender, L. R., and J. C. K. Kim. (1988) "The Design and Implementation of Disaster Mitigation Policy." In *Managing Disaster: Strategies and Policy Perspective*. Ed. L. K. Comfort. 67–85. Durham N.C.: Duke University Press.

Energimyndigheten [State Energy Authority]. (1999) *Nya Zeelands elmarknad: Regelverket revideras efter det allvarliga elavbrottet i Auckland 1998* [The New Zealand Electricity Market: Regulations Scrutinised Following the Serious Power Outage in Auckland 1998]. Eskilstuna: Energimyndigheten.

Energy Information Administration (EIA). "Country Energy Data Report: New Zealand," U.S. Department of Energy (DOE) Statistical Office. Available at: http://www.eia.doe.gov/emeu/world/country/cntry_NZ.html

Energy Information Administration (EIA). "Country Energy Data Report: Sweden," U.S. Department of Energy (DOE) Statistical Office. Available at: http://www.eia.doe.gov/emeu/world/country/cntry_SW.html

Energy-Wise News. (1998) "Power Crisis Businesses Consider Teleworking." Ed. Cathy Sheehan. Issue 58. June/July. Available at: http://www.energywise.co.nz/58june98

Farnham, B. (1997) *Roosevelt and the Munich Crisis: A Study of Political Decision-Making*. Princeton, N.J.: Princeton University Press.

Fischer, Frank. (1990) *Technocracy and the Politics of Expertise*. Newbury Park, Calif.: Sage.

Fiske, S., and S. Taylor. (1991). *Social Cognition*. 2nd ed. New York: McGraw-Hill.

Fitzsimons, J. (1998) "Parliamentary Session of the Government of New Zealand" Hansard (02/24/98). Available at: http//rangi.knowledge-basket.co.nz/hansard/han/text/1998/02/24-019.html

Flin, R. (1996) *Sitting in the Hot Seat: Leaders and Teams for Critical Incident Management*. Chichester, U.K.: Wiley.

Fricker, A. (1998). "The Ecological Footprint of New Zealand as a Step Towards Sustainability." *Futures*, 30(6):559–67.

Gallagher, M. (1998) "The Political Impact of Electoral System Change in Japan and New Zealand 1996." *Party Politics*, 4(2):203–28.

Gardiner, J. (1998) "Downtown Shutdown: City Declared No-Go Zone." *The New Zealand Herald* (02/23/98).

Gardiner, J. (1998) "Shudders in the Suburbs: Homes Beyond Central City Hit by Electricity Rationing." *The New Zealand Herald* (02/24/98).

Gardiner, J. (1998) "Into an Autumn of Gloom: Seven Weeks without Full Power." *The New Zealand Herald* (03/05/98).

Gardiner, J. (1998) "Insurance Finding Likely to Help Powerless." *The New Zealand Herald* (03/31/98).

Gardiner, J., and Eugene Bingham. (1998) "Erratic Supply Hits Back-to-City Hopes." *The New Zealand Herald* (03/09/98).

Gardner, H. (1987) *The Mind's New Science*. New York: Basic Books.

Gardner, H. (1995) *Leading Minds: An Anatomy of Leadership*. New York: Basic Books.

Gascoigne, T., and Metcalfe, J. (1997) "Incentives and Impediments to Scientists Communicating through the Media." *Science Communication: An Interdisciplinary Social Science Journal*, 18(3):265–82.

Gaynor, B. (1999) "Contact Needs Bigger Share." New Zealand Herald Online Edition (10/04/99). Available: http://www.nzherald.co.nz

Gelb, L., and Betts, R. (1979) *The Irony of Vietnam: The System Worked*. Washington, D.C.: Brookings Institute.

George, A. (1980) *Presidential Decision-Making in Foreign Policy: The Effective Use of Information and Advice*. Boulder, Colo.: Westview Press.

George, A. (1986) "The Impact of Crisis-Induced Stress on Decision Making." *The Medical Implications of Nuclear War*. Washington, D.C.: National Academy Press.

George, A. (1991) "The Cuban Missile Crisis." In *Avoiding War: Problems of Crisis Management.* Ed. Alexander George, 222–68. Boulder, Colo.: Westview Press.

George, A. (1993) *Bridging the Gap: Theory and Practice in Foreign Policy*. Washington, D.C.: United States Institute of Peace Press.

George, A., and Smoke, R. (1974) *Deterrence in American Foreign Policy: Theory and Practice*. New York: Columbia University Press.

Gerard, J. (1997) "New Zealand: The Reformed Economy." Address to the Semi-Annual General Meeting of Credit Union Central of British Columbia. Vancouver, 21 November, 1997. Ottawa: High Commission of New Zealand.

Gillett, G. (1998) "Justice and Health Care in a Caring Society." *The British Medical Journal,* 316(7150): 53–54.

Gillon, G. (1998) "Parliamentary Session of the Government of New Zealand" Hansard (02/24/98). Available at: http//rangi.knowledge-basket.co.nz/hansard/han/text/1998/02/24-019.html

Government of New Zealand. (1983) Civil Defence Act of 1983 (no.46). Wellington.

Government of New Zealand. (1988) Civil Defence Amendment Act of 1988 (no.54). Wellington.

Governor's Commission on the Los Angeles Riots. (1965) *Violence in the City: An End or a Beginning?* Los Angeles: The Commission, 1965.

Graham, S. (2002) "Bridging Urban Digital Divides? Urban Polarisation and Information and Communications Technologies (ICTs)." *Urban Studies* 39, (1):33–56.

Green, D., and Shapiro, I. (1994) *Pathologies of Rational Choice Theory: A Critique of Applications in Political Science*. New Haven, Conn.: Yale University Press.

Grid Security Policy. (1998) "A Joint Electricity Industry Project." *Update*, 3/98. Available: http://www.gsp.co.nz

Grönvall, J. (1998) "Galna ko-krisen, 1996" [The Mad-Cow Crisis, 1996]. In *Crisis and Internationalisation: Eight Crises Studied From a Cognitive-Institutional Perspective.* Ed. Eric Stern and Fredrik Bynander. Stockholm: ÖCB [The Swedish Agency for Civil Emergency Planning].

Habermas, J. (1973/1976) *Legitimation Crisis*. Trans. Thomas McCarthy. London: Heinemann.

Hagstrom, A.-Z., and Sundelius, B. (2001) *Krishantering på göteborgska: En studie av brandkatastrofen den 29-30 oktober 1998* [*Crisis Management the Gothenburg Way: A Study of the Fire Catastrophe on October 29–30, 1998*], Vol. 15, Stockholm: ÖCB [Swedish Agency for Civil Emergency Planning].

Haimes, Y. Y. (2002) "Risk of Terrorism to Cyber-Physical and Organizational-Societal Infrastructures." *Public Works Management & Policy* 6, (4):231–40(10).

Halbert, D. J. (2002) "Citizenship, Pluralism, and Modern Public Space." *Innovation: The European Journal of Social Sciences* 15, (1):33–42(10).

Halperin, M. (1974) *Bureaucratic Politics and Foreign Policy*. Washington, D.C.: Brookings Institute

Haney, P. J. (1997) *Organizing for Foreign Policy Crises: Presidents, Advisers, and the Management of Decision Making*. Ann Arbor: University of Michigan Press.

Hansén, D., and E. Stern. (2001) "From Crisis to Trauma: The Palme Assasination Case." In *Managing Crises: Threats, Dilemmas, Opportunities*. Ed. U. Rosenthal, A. Boin, and L. K. Comfort. Springfield, Ill.: Charles C. Thomas

Hansén, D. (1998) "The Occupation of the West-German Embassy in Stockholm on April 24, 1975 and Its Management." Unpublished manuscript. Stockholm: Swedish Institute of International Affairs.

Hardie Boys, Sir M. (1998) "The Constitutional Challenges of MMP: A Magical Demystification Tour." Speech given to the Institute of Policy Studies, Victoria University of Wellington Council Chamber, on 3 December 1998. Available at: http://www.gov-gen.govt.nz/speeches/index.html

Harris, P. (1997) "The Electoral Commission." In *New Zealand Politics in Transition*. Ed. Raymond Miller. Auckland: Oxford University Press.

't Hart, P. (1993) "Symbols, Rituals and Power: The Lost Dimensions of Crisis Management." *Journal of Contingencies and Crisis Management,* 1(1):36–50.

't Hart, P. (1999) Workshop on "Crisis Management in the Metropolis." International Conference on the Future of European Crisis Management in The Hague, 7–9 November.

't Hart, P., U. Rosenthal, and Boin A. (1999) "From Crisis to 'Normalcy': The Long Shadow of 'Post-Crisis' Politics." In *From Crises to Contingencies: A Global Perspective*. Ed. Uriel Rosenthal, Louise Comfort, and Arjen Boin.

't Hart, P., Stern, E. and Sundelius, B., eds. (1997) *Beyond Groupthink Political Group Dynamics and Foreign Policy-Making*. Ann Arbor: University of Michigan Press.

Hayward, B. (1997) "Public Participation." In *New Zealand Politics in Transition*. Ed. Raymond Miller. Auckland: Oxford University Press.

Heaps, B. (1998) Electricity System Update: February 11, 1998. New Zealand: Media Release Archive Transpower. Available: http://www.transpower.co.nz/news/media/19980212electricity_html

Hellberg, A. and Jörle, A. (1984) *Ubåt 137: Tio dagar som skakade Sverige*. [Submarine 137: Ten days that shook Sweden]. Stockholm: Atlantis.

Henry, R. A. (2000) *The Complete Guide to Crisis and Risk Communications . . . You'd Better Have a Hose if You Want to Put Out the Fire: Professional Tips, Tactics, Dos, Don'ts and Case Histories*. Windsor, Calif.: Gollywobbler Productions.

Hermann, C. (1963) "Some Consequences of Crisis Which Limit the Viability of Organizations." *Administrative Science Quarterly*, 8:61–82.

Hermann, M. G., and J. D. Hagan. (1998) "International Decision Making: Leadership Matters." *Foreign Policy* 110 (spring): 124–37. Special Edition: *Frontiers of Knowledge: The State of the Art in World Affairs*.

Hermann, M. G., C. F. Hermann, and J. Hagan. (1987) "How Decision Units Shape Foreign Policy Behavior." In *New Directions in the Study of Foreign Policy*. Ed. C. F. Hermann, C. W. Kegley, and J. N. Roseneau, 309–338. Boston: Allen and Unwin.

Hermann, M. (1980) "Explaining Foreign Policy Behavior Using Personal Characteristics of Political Leaders." *International Studies Quarterly* 24:7–46.

Hermann, M. (1979) "Indicators of Stress in Policy Makers During Foreign Policy Crises" *Political Psychology* 1(1):27–46.

Higgins, E., and Bargh, J. (1987) "Social Cognition and Social Perception." *Annual Review of Psychology,* 38:369–425.

Hodgson, P. (1998) "Parliamentary Session of the Government of New Zealand" Hansard (02/24/98). Online. Available: http://rangi. knowledgebasket.co.nz/hansard/han/text/1998/02/24-020.html

Hollis, M., and Smith, S. (1990) *Explaining and Understanding International Relations.* Oxford, U.K.: Clarendon Press.

Holsti, O. (1989) "Crisis Decision Making," In *Behavior, Society, and Nuclear War*. Ed. Philip Tetlock et al,. pp 8–84. New York: Oxford University Press.

Hoyt, P. (1997) "The Political Manipulation of Group Composition: Engineering the Decision Context." *Political Psychology*, 18:771–90.

Hult, K. (1993) "Advising the President." In *Researching the Presidency*. Ed. George Edwards, John Kessel and Bert Rockman, 111–59. Pittsburgh, Penn.: University of Pittsburgh Press.

Human Resources Group: Auckland City Council. (1998) "General News: Monthly Group Managers Meeting: Power Usage Problems—Auckland CBD (02/18/98)" Insite. Auckland, New Zealand: Auckland City Council.

Hyde, T. (1998) "Power Play." *Metro*, 202: 31–36.

Häggroth, S., K. Kronvall, C. Riberdahl, and K. Rudebeck. (1993) *Swedish Local Government: Traditions and Reforms*. Stockholm: Swedish Institute.

IEA. (1999) *Key World Energy Statistics from the IEA: 1999 Edition.* Paris: International Energy Agency.

Impact. (1998) "New Zealand's New Direction in Emergency Management—Why Are We Doing this Anyway?" *Impact: The Newsletter of Emergency Management in New Zealand*, (autumn), Wellington: Ministry for Emergency Management.

Integral Energy. (1998) "Inquiry into the Auckland Power Supply Failure—Distribution Planning Review." Auckland: Integral Energy.

Jajn, D. (1998) "Environmental Performance and Policy Regimes: Explaining Variations in 18 OECD Countries." *Policy Sciences,* 31:101–13.

Jacobs, B. (1997) "Urban Crisis: Complexity and Risk—an Editorial Statement." *Journal of Contingencies and Crisis Management*, 5(3):127–30.

Janis, I. (1982) *Groupthink: Psychological Studies of Decisions and Fiascoes.* 2nd ed. Boston: Houghton Mifflin.

Janis, I. (1989) *Crucial Decisions: Leadership in Policymaking and Crisis Management*. New York: Free Press.

Janis, I., and Mann, L. (1977) *Decision Making: A Psychological Analysis of Conflict, Choice, and Commitment*. New York: Free Press.

Jervis, R. (1976) *Perceptions and Misperceptions in International Politics*. Princeton, N.J.: Princeton University Press.

Jervis, R. (1997) *System Effects: Complexity in Political and Social Life*. Princeton, N.J.: Princeton University Press.

Kärde, T. (1999) "The Holland Tunnel Crisis, 1997" Unpublished report. Crisis Management Europe—National Crisis Management from an International Perspective. Stockholm.

Katzenstein, P. (ed.) (1996) *The Culture of National Security: Norms and Identity in World Politics*. New York: Columbia University Press.

Kearney, R. C. and C. Scavo. (2001) "Reinventing Government in Reformed Municipalities: Manager, Mayor, and Council Actions." *Urban Affairs Review*, September, 37(1):43–66.

Keeney, R. (1992) *Value Focused Thinking*. Cambridge, Mass.: Harvard University Press.

Khong, Y. (1992) *Analogies at War: Korea, Munich, Dien Bien Phu and the Vietnam Decisions of 1965*. Princeton, N.J.: Princeton University Press.

Kingdon, J. (1984) *Agendas, Alternatives, and Public Policies*. Boston: Little, Brown.

Krebs, T. B., and J. P. Pelissero. (2001) "Fund-Raising Coalitions in Mayoral Campaigns." *Urban Affairs Review* 37(1):67–84.

Ladley, A. (1997) "The Head of the State: The Crown, the Queen and the Governor-General." In *New Zealand Politics in Transition*. Ed. Raymond Miller. Auckland: Oxford University Press.

Lagadec, P. (1997) "Learning Processes for Crisis Management in Complex Organizations." *Journal of Contingencies and Crisis Management,* 5(1):24–31.

La Porte, T., and P. Consolini. (1991) "Working in Practice But Not in Theory: Theoretical Challengs of High-Reliability Organizations," *Journal of Contingencies and Crisis Management*, 2(4):221–27.

Larson, D. (1994) "The Role of Belief Systems and Schemas in Foreign Policy Decision-Making." *Political Psychology*, 15(1):17–33.

Larson, D. W. (1985) *The Origins of Containment: A Psychological Explanation*. Princeton, N.J.: Princeton University Press.

Lebow, R., and J. Stein. (1994) *We All Lost the Cold War*. Princeton, N.J.: Princeton University Press.

Lebow, R. (1998) "Beyond Parsimony: Rethinking Coercieve Bargaining." *European Journal of International Relations*, 4(1):31–66.

Lebow, R. (1981) *Between Peace and War: The Nature of International Crisis*. London: Johns Hopkins University Press.

Lee, S. (1998) "Parliamentary Session of the Government of New Zealand." Hansard (02/24/98). Online. Available: http://rangi.knowledgebasket.co.nz/hansard/han/text/1998/02/24-020.html

Levine, S., and N. Roberts. (1997a) "MMP: The Decision." In *New Zealand Politics in Transition*. Ed. Raymond Miller. Auckland: Oxford University Press.

Levine, S., and N. Roberts. (1997b) "The 1996 General Election." In *New Zealand Politics in Transition*. Ed. Raymond Miller. Auckland: Oxford University Press.

Levy, J. (1994) "Learning and Foreign Policy: Sweeping a Conceptual Minefield." *International Organization,* 48(2):279–312

Levy, J. (1991) "The Role of Crisis Management in the Outbreak of World War I." In *Avoiding War: Problems of Crisis Management*. Ed. A. L. George, 62–102. Boulder, Colo.: Westview Press.

Lindblom, C. (1990) *Inquiry and Change*. New Haven, Conn.: Yale University Press.

Lifelines. (1997) *Auckland Engineering Lifelines Project: Stage One Report 1997*. Auckland: RSON Group.

Lifelines. (1998) *Risk Management Seminar: Looking Forward from the Auckland Power Crisis*. Auckland: RSON Group.

Linqvist, S., P. Wikberg, J. Trued, and B. Bergh. (1999) *Applied Training with an Organisation Test in the Rescue Centre of Stockholm [1999–05–27]*: *Valuation of the Command Organisations for Local Government, Rescue Service and Police*. Stockholm: National Defence Research Establishment (FOA). R-99-01161-990-SE.

Local Government New Zealand. (LGNZ) (1999) "Local Democracy 2000." Available: http://www.lgnz.co.nz/

Long, John. (1996) Speaker, Presentation at International Conference on Crisis Management in Skåvsjöholm, Sweden.

Lorrain, D. (2001) "Gig@city: The Rise of Technological Networks in Daily Life." *Journal of Urban Technology*, 1 December 2001, 8(3):1–20.

Lund, M. S. (1996) *Preventing Violent Conflicts: A Strategy for Preventive Diplomacy*. Washington, D.C.: United States Institute of Peace Press.

March, J., and J. Olsen. (1989) *Rediscovering Institutions: The Organizational Basis of Politics*. New York: Free Press.

March, J. (1991) "Exploration and Exploitation in Organizational Learning." *Organizational Science,* 2(1):71–87.

Marketplace Company, (M-co). (1999) "Hugh Rennie Appointed Chair of M-co" Press Release, 3 June, 1999. Available at: http://www.m-co.co.nz

Markus, H., and R. Zajonc. (1985) "The Cognitive Perspective in Social Psychology." In *The Handbook of Social Psychology*. 3rd ed. Ed. L. Gardiner and R. Zajonc, 250–60. Reading, Mass.: Addison-Wesley.

Matthews, C. (1988) *Hardball: How Politics is Played, Told by One Who Knows the Game*. New York: Summit Books.

Maurino, D. E., et al. (1995) *Beyond Aviation Human Factors: Safety and High Technology Systems*. Hants, England: Avebury Aviation.

McClintock, B. (1998) "Whatever Happened to New Zealand? The Great Capitalist Restoration Reconsidered." *Journal of Economic Issues,* 32(2):497–503.

McKinlay, P. (1998) Case Study on the Auckland Regional Services Trust (ARST), a subset in the Project on Public Ownership and Accountability. Request from author: http://www.mdl.co.nz

McQueen, A. (1998) "New Zealand: Physical and Social Geography" *The Far East and Australasia*. 13th ed. London: Europa Publications.

Meertens, J. V. (2000) "The Bijlmer Disaster: Does a Public Inquiry Really Help Us Learn from a Disaster?" Raamsdonksveer: International Centre for Emergency Techniques (ICET), The Netherlands.

Mercury Energy Limited (1997) *Mercury Energy Limited's 1997 Annual Report: Positioned for the Future*. Auckland: Mercury Energy Limited.

Mercury Energy Limited (1998) Press Release (02/09/98-11 A.M.).

Mercury Energy Limited, (1998) Press Release (02/09/98-12 P.M.).

Mercury Energy Limited (1998) Press Release (02/10/98-11 A.M.).

Mercury Energy Limited (1998) Press Release (02/10/98-12 P.M.).

Mercury Energy Limited (1998) Press Release (02/10/98-4 P.M.).

Mercury Energy Limited (1998) Press Release (02/11/98).

Mercury Energy Limited (1998) Press Release (02/12/98).

Mercury Energy Limited (1998) Press Release (02/13/98).

Mercury Energy Limited (1998) Press Release (02/15/98).

Mercury Energy Limited (1998) Press Release (02/17/98).

Mercury Energy Limited (1998) Press Release (02/19/98).

Mercury Energy Limited (1998) Press Release (02/20/98-11 A.M.).

Mercury Energy Limited (1998) Press Release (02/20/98-2.30 P.M.).

Mercury Energy Limited (1998) Press Release (02/21/98-10.30 A.M.).

Mercury Energy Limited (1998) Press Release (02/21/98-12 P.M.).

Mercury Energy Limited (1998) Press Release (02/23/98).

Mercury Energy Limited (1998) Press Release (02/23/98b).

Mercury Energy Limited (1998) Press Release (02/24/98).

Mercury Energy Limited (1998) Press Release (02/24/98-1:30 P.M.).

Mercury Energy Limited (1998) Press Release (03/01/98).

Miller, R. (1997) "Preparing for MMP: 1993–1996." In *New Zealand Politics in Transition*. Ed. Raymond Miller. Auckland: Oxford University Press

Miller, R., and J. Malinowski. (1994) *Power System Operation.* 3rd ed. New York: McGraw-Hill.

Miller, A. H. (2001) "The Los Angeles Riots: A Study in Crisis Paralysis." *Journal of Contingencies and Crisis Management*, 9(4):189–99.

Miller, J. (2002) "Affirming Flames: Debriefing Survivors of the World Trade Center Attack." *Brief Treatment and Crisis Intervention*, March, 2(1):85–94(10).

Ministry of Civil Defence, New Zealand. (MoCD) (1998) The Auckland Supply Disruption 1998: Review of Emergency Management Response. June 1998 (E22/1).

Minister of Civil Defence with the Assistance of the Historical Branch: Department of Internal Affairs (MoCD). (1990) *Civil Defence in New Zealand: A Short History*. Wellington: Minister of Civil Defence.

Ministry of Civil Defence, New Zealand (MoCD). (1993) "The Civil Defence Act of 1983. In National Civil Defence Plan-Part Two: Disaster Recovery. Amendment 4(3/92)." Recovery Plan: Natural Disasters and Emergencies Within New Zealand. Online. Available: http://www.mocd.govt.nz/

Ministry of Civil Defence, New Zealand (MoCD). (1993) "Introduction: Amendment 10 (6/96)." National Civil Defence Plan. Online. Available at: http://www.mocd.govt.nz/

Ministry of Civil Defence, New Zealand (MoCD). (1998) Fax "Draft Welfare Accommodation Plan—Auckland CBD Energy Crisis Emergency Management Plan." Annex A the Auckland CBD Energy Crisis Emergency Management Plan.

Ministry of Civil Defence (MoCD)/Ministry for Emergency Management. (2000) "Ministry for Emergency Management: Who We Are: The Ministry: The Civil Defence Act." Online. Available at: http://www.mem.govt.nz/memwebsite.nsf

Ministry of Commerce, New Zealand. (1998) *Auckland Power Supply Failure 1998: The Report of the Ministerial Inquiry into the Auckland Power Supply Failure*. Wellington: Ministry of Commerce of New Zealand.

Ministry of Commerce, New Zealand. (1999b) Chronology of New Zealand Electricity Reform. Online. Available at: http://www.moc.govt.nz//ran/empg/chronology/index.html

Ministry of Commerce, New Zealand. (1999a) The 1999 Electricity Regulations: Information Disclosure. Online. Available at: http://www.moc.govt.nz/ers/inf-disc/eidr1999/eidr1999.html

Ministry of Defence, New Zealand. (1983) The Civil Defence Act of 1983 No. 46, Section 2: Interpretation. Online. Available at: http://www.mocd.govt.nz/

Ministry for Emergency Management. (2000) "Preparing Civil Defence Emergency Expenditure Responsibility Claims: Guidelines for Local Government," March. Wellington: Ministry for Emergency Management. Available at: http://www.mem.govt.nz/memwebsite.nsf

Ministry for Emergency Management. (1999) "Resilient New Zealand: Realising the Potential." Briefing for the Ministry for Emergency Management. Online. Available at: http://www.mocd.govt.nz

Mintz, A. (1997) "Foreign Policy Decisionmaking: Bridging the Gap Between the Cognitive Psychology and Rational Actor 'School.'" In *Decision-Making on War and Peace: The Cognitive-Rational Debate*. Ed. N. Geva and A. Mintz, 1–10. Boulder, Colo.: Lynne Rienner.

Mitroff, I. I., C. M. Pearson, and L. K. Harrington. (1996) *The Essential Guide to Managing Corporate Crises: A Step-By-Step Handbook for Surviving Major Catastrophes*. Oxford, U.K.: Oxford University Press.

Moens, A. (1991) "President Carter's Advisors and the Fall of the Shah." *Political Science Quarterly*, 106(2):211–237.

Morello-Frosch, R., M. Pastor, and J. Sadd. (2001) "Environmental Justice and Southern California's 'Riskscape': The Distribution of Air Toxics Exposures and Health Risks Among Diverse Communities." *Urban Affairs Review*, March, 36, (4): 551–78.

Morgenthau, H. (1948) *Politics Among Nations: The Struggle for Power and Peace*. New York: Knopf.

Mulgan, R. (1997) "Parliament: Composition and Functions." In *New Zealand Politics in Transition*. Ed. Raymond Miller. Auckland: Oxford University Press.

Mulgan, R. (1999) "Have New Zealand's Political Experiments Increased Public Accountability? Discussion Paper No. 59" Auckland: Department of Public Policy, Australian National University. Online. Available at: http://www.anu.edu.au/pubpol/people/docs/dp_59.html

Nagel, J. H. (1998) "Social Choice in a Pluralitarian Democracy: The Politics of Market Liberalization in New Zealand." *British Journal of Political Science*, 28: 223–67, Cambridge, U.K.: Cambridge University Press.

Nelson, A. L., K. P. Schwirian, and P. M. Schwirian. (1998) "Social and Economic Distress in Large Cities, 1970–1990: A Test of the Urban Crisis Thesis." *Social Science Research*, December, 27(4): 410–31.

Neustadt, R. (1990) *Presidential Power and the Modern Presidents: The Politics of Leadership from Roosevelt to Reagan*. New York: Free Press.

Neustadt, R., and E. May. (1986) *Thinking in Time: The Use of History for Decision Makers*. New York: Free Press.

Newlove, L. (1998) "Ice Storm in Eastern Canada, 1998" Unpublished report. Crisis Management Europe—National Crisis Management from an International Perspective. Stockholm.

New Zealand Alliance Party Press Release. (1998) "Russel McVeagh's Role Targeted: Newsroom Politics." Published February 25, 1998. Online. July 11, 1998. Available at: http://www.newsroom.co.nz/stories/PO9802/SOO371.html

The New Zealand Dairy Exporter. (1998) "The Exports Must Get Through—and They Do." *The New Zealand Dairy Exporter*. March:70.

New Zealand Fire Services. (1998) "Operational and Civil Defence Emergency Plans—Operation Energy." Auckland: New Zealand Fire Services.

New Zealand Government Online. (1999a) "The New Zealand Government" Online. Available at: http://www.govt.nz/nzgovt/nzgovt.php3

New Zealand Government Online. (1999b) "Judiciary and Courts" Online. Available at: http://www.govt.nz/nzgovt/courts.php3

The New Zealand Herald ("In Short"). (1998) "Mercury Disguises Workers." *The New Zealand Herald*. O2/26/98.

New Zealand Department of Civil Defence and Emergency Management. (1998) "A Risky Business," Inform (06/11/98). New Zealand: Department of Civil Defence and Emergency Management. Available at: http://mocd.govt.nz

Nisbett, R., and L. Ross. (1980) *Human Inference: Strategies and Shortcomings of Human Judgement*. Englewood Cliffs, N.J.: Prentice Hall.

Nohrstedt, S. A. (1991) "The Information Crisis in Sweden after Chernobyl." *Mass Media, Culture, and Society*, 13: 477–97.

Nohrstedt, D. (2000) *Managing Crisis Abroad: The Brolin Kidnapping. Stockholm: ÖCB* [Swedish Agency for Civil Emergency Planning].

Nohrstedt, D., and B. Sundelius. (1999) *Introduction to Crisis Management in Estonia: Case Studies and Comparative Perspectives*. Ed. E. Stern and D. Nohrstedt. Stockholm: ÖCB [Swedish Agency for Civil Emergency Planning].

Nohrstedt, S. A., and A. Tassew. (1993) *Communication and Crisis: An Inventory of Current Research*. Psykologiskt försvar, Report no.163:1.

Nordlund, R. (1994) *A Triangle Drama: Authorities, Citizens and Media in Crisis*. Psykologiskt försvar, Report no. 136:b.

Norris, D. F. (2001) "Whither Metropolitan Governance?" *Urban Affairs Review*, March, 36(4): 532–50.

North, P., D. Valler, and A. Wood. (2001) "Talking Business: An Actor-Centred Analysis of Business Agendas for Local Economic Development." *International Journal of Urban and Regional Research*, December, 25(4): 830–46.

Norton, J. (1999) Foreword. "Resilient New Zealand: Realising the Potential" Briefing for the Minister Responsible for the Ministry of Emergency Management. December. Online. Available at: http://www.mem.govt.nz/memwebsite.nsf

OECD. (1996) *OECD Economic Surveys 1996–1997: Sweden*. Paris: OECD.

Officer, R. (1996) "Foreword." In *Transforming Government Enterprises: Managing Radical Organisational Change in Deregulated Environments*. Ed. B. Spicer, D. Emmanvel, M. Powell. St. Leonards, Australia: Centre for Independent Studies.

Paige, G. (1968) *The Korean Decision*. New York: Free Press.

Pearce, D. D. (1995) *Wary Partners: Diplomats and the Media*. Washington, D.C.: Congressional Quarterly Press.

Pearsall, J. and Trumble, B., eds. (1996) *The Oxford English Reference Dictionary*. 2nd ed. Oxford, U.K.: Oxford University Press.

Pennebaker, J. W. and T. C. Lay. (2002) "Language Use and Personality during Crises: Analyses of Mayor Rudolph Giuliani's Press Conferences." *Journal of Research in Personality*, June, 36(3): 271–82(12).

Perrow, C. (1994) "The Limits of Safety: The Enhancement of a Theory of Accidents." In *The Journal of Contingencies and Crisis Management*, 2(4):213–20.

Perrow, C. (1999) *Normal Accidents: Living with High-Risk Technologies*. Princeton, N.J.: Princeton University Press.

Peters, W. "Parliamentary Session of the Government of New Zealand" Hansard (02/24/98). Available at: http//rangi.knowledge-basket.co.nz/hansard/han/text/1998/02/24-019.html

Pfeffer, J. (1981) *Power in Organizations*. Marshfield, Mass.: Pittman.

Porfiriev, B. (2002) "Managing the Alleviation of the 1995 Neftegorsk Earthquake." In *Crisis Management in Russia: Overcoming Institutional Rigidity and Resource Constraints*. Ed. Boris Porfiriev and Lina Svedin, 17: 131–72. Stockholm: ÖCB [Swedish Agency for Civil Emergency Planning].

Post, J. (1991) "The Impact of Crisis-Induced Stress of Decision Makers" In *Avoiding War*. Ed. A. George. Boulder, Colo.: Westview Press.

Powell, P., and P. DiMaggio, eds. (1991) *The New Institutionalism in Organizational Analysis*. Chicago: University of Chicago Press.

Power Engineering. (1999) "EIA: Industrial Power Cheapest in 17 Years." 103 (2): 14.

The Press–New Zealand News (1998) "Death Delays Crisis Report." (07/15/98).

Pressman, J., and A. Wildavsky. (1984) *Implementation: How Great Expectations in Washington are Dashed in Oakland*. Berkeley: University of California Press.

Purkitt, H. (1992) "Political Decision Making in Small Groups." In *Political Psychology and Foreign Policy*. Ed. E. Singer and V. Hudson. 219–45. Boulder, Colo.: Westview Press.

Quarantelli, E. L. ed. (1998) *What is a Disaster? Perspectives on the Question*. London: Routledge.

Rand Merchant Bank Australia (RMBA). (1999) "Rand Merchant Bank Acquires M-co." Available at: http://www.rmb.com.au

Regester, M., and J. Larkin. (1998) *Risk Issues and Crisis Management: A Casebook of Best Practice*. London: Institute of Public Relations.

Rein, M., and D. Schön. (1991) "Frame-reflective Policy Discourse." In *Social Sciences and Modern States*. Ed. P. Wagner et al. 262–289. Cambridge, U.K.: Cambridge University Press.

Reynolds, M. (02/23/98) "The Man Who Lost the Battle to Keep the City's Lights On." *The New Zealand Herald*. Online. Available at: http://www.nzherald.co.nz

Reynolds, M. (01/03/99) "Impartial Eye Vital Say Power Watchers." *The New Zealand Herald*. Online. Available at: http://www.nzherald.co.nz

Reynolds, M. (12/06/99) "Consumers May Yet Cash In: Customers Could Benefit if Electricity Wholesale Prices Remain Below Pre-April 1st Levels." *The New Zealand Herald Online Edition*. Online. Available at: http://www.nzherald.co.nz

Reynolds, M. (16/06/99). "Vector Profits Hit by Power Contract Costs." *The New Zealand Herald Online Edition*, Online. Available at: http://www.nzherald.co.nz

Rosati, J. A. (1981) "Developing a Systematic Decision-Making Framework: Bureaucratic Politics in Perspective." *World Politics* 33(2):234–252.

Rose, R. (1994) *Inheritance in Public Policy: Change Without Choice in Britain*. New Haven, Conn.: Yale University Press.

Rosenthal, U., A. Boin, and L. K. Comfort, eds. (2001) *Managing Crises: Threats, Dilemmas, Opportunities*. Springfield, Ill: Charles C. Thomas.

Rosenthal, U., M. Charles, and P. 't Hart, eds. (1989) *Coping With Crises: The Management of Disasters, Riots, and Terrorism*. Springfield, Ill.: Charles C. Thomas.

Rosenthal, U. and P. 't Hart, (1991) "Experts and Decisionmakers in Crisis Situations." *Knowledge* 12:350–73.

Rosenthal, U., P. 't Hart, and A. Kouzmin. (1991) "The Bureau-Politics of Crisis Management." *Public Administration* 69:211–33.

Rosenthal, U. and P. 't Hart, (eds.). (1998) *Flood Response and Crisis Management in Western Europe: A Comparative Analysis*. Berlin: Springer Verlag.

Rosenthal, U., P. 't Hart, M. van Duin, A. Boin, M. Kron, M. Otten, and W. Overdijk. (1994) *Complexity in Urban Crisis Management: Amsterdam's Response to the Bijlmer Air Disaster*. London: James and James.

Ross, M. (1998) "Few Blackout Plans in Auckland." *Chartered Accountants Journal of New Zealand,* April: 44.

Ruth, J. (1998). "Power Plays." *Independent Review,* 279(7):21.

Sabatier, P. (1993) "Top-Down and Bottom-Up Approaches to Implementation Research." In *The Policy Process: A Reader*. Ed. M. Hill. New York: Harvester Wheatsheaf, 26–93.

Sagan, S. (1993) *The Limits of Safety: Organizations, Accidents and Nuclear Weapons*. Princeton, N.J.: Princeton University Press.

Sampson, M. (1987) "Cultural Influences On Foreign Policy," In *New Directions in the Study of Foreign Policy*. Ed. C. Hermann, Kegley, and J. Rosenau, 384–405. Boston: Allen and Unwin.

Security Development Working Group. (1998). "Matrix of Security Issues. Discussion paper." 28 January. Available at: http://www.gsp.co.nz

Shafi, S. S. (1994) "Plague Epidemic: Part of the Urban Crisis." *Nagarlok*, 26(4):53.

Sinclair, Knight, Mertz (SKM)/Leyland Consultants (2000) "Electricity Supply and Demand to 2015." A "Leyline Story." Available at: http://ourworld.compuserve.com/homepages/Leyland_Consultants/leyline7.html

Singer, M. (1994) "AIDS and the Health Crisis of the U.S. Urban Poor: The Perspective of Critical Medical Anthropology." *Social Science & Medicine*, 39(7):931.

Sjöblom, G. (1993) "Some Critical Remarks on March and Olsen's Rediscovering Institutions." *Journal of Theoretical Politics* 5(3):397–407.

Skinner, P. G., A. J. Schear, and S. S. Katz (2002) "Leasing Issues and Provisions in the Face of Catastrophic Loss: A Post-11 September Discussion." *Journal of Corporate Real Estate*, April, 4(2):112–21(10).

Smith, H. (1988) *The Power Game: How Washington Works*. New York: Ballantine Books.

Smith, S., (1985) "The Hostage Rescue Mission." In *Foreign Policy Implementation*. Ed. S. Smith and M. Clarke. London: Allen and Unwin.

Smith, S., and M. Clarke, eds. (1985) *Foreign Policy Implementation*. London: Allen and Unwin.

Snook, S. (1993) *The Limits of Safety: Organizations, Accidents, and Nuclear Weapons*. Princeton, N.J.: Princeton University Press.

Snyder, R., H. Bruck, and B. Sapin, eds. (1963) *Foreign Policy Decision Making*. New York: Free Press.

Soltan, K. (1998) "Institutions as Products of Politics." In *Institutions and Social Order*. Ed. K. Soltan, E. Uslander and V. Haufler. 45–64. Ann Arbor: University of Michigan Press.

Soltan, K., E. Uslander, and V. Haufler. (1998) "New Institutionalism: Institutions and Social Order." in *Institutions and Social Order*. Ed. K. Soltan, E. Uslander, and V. Haufler. 3–14. Ann Arbor: University of Michigan Press.

SOU (Statens Offentliga Utredningar). 1996:189 (1996) "Efter Estonia: Rapport av Utredningen för vägledning efter Estoniakatastrofen." Stockholm: Fritzes-Offentliga Publikationer.

Speden, G. (1998) "Power Crisis Spur for Reform." *The Independent*. February 25.

Spicer, B., D. Emanuel, and M. Powell. (1996) "Transforming Government Enterprises: Managing Radical Organisational Change." Deregulated Environments. St. Leonards, Australia: Centre for Independent Studies.

Staples, R. (1994) "The Urban Crisis Revisited: A Barometer of the Future in Race Relations." *International Journal of Contemporary Sociology*, 31(2):183.

Statistics New Zealand. (1999) "New Zealand in Profile 1999" (brochure). Available at: http://www.stats.govt.nz

Staw, B., L. Sandelands, and J. Dutton. (1981) "Threat-Rigidity Effects in Organizational Behavior: A Multi-level Analysis." *Administrative Science Quarterly*. 26:501–24.

Steen, G. et al., (1987) *Beredskap efter Tjernobyl: Slutrapport*. Stockholm: Statens Haverikommission.

Stehr, N. (1991) "The Power of Scientific Knowledge and Its Limits." *Canadian Review of Sociology and Anthropology*. 28(4):460–481.

Stein, J., and D. Welch. (1997) "Rational and Psychological Approaches to the Study of International Conflict" In *Decision-Making on War and Peace: The Cognitive-Rational Debate*. Ed. N. Geva and A. Mintz. 51–80. Boulder, Colo.: Lenne Rienner.

Steinbruner, J. (2000) *Principles of Global Security*. Washington, D.C.: Brookings Institute.

Steinbruner, J. (1974) *The Cybernetic Theory of Decision*. Princeton, N.J.: Princeton University Press.

Steitz, D., and T. Tyson. (1998) "NASA Helps 'Hot' Cities Cool Down," press release. Marshall Space Flight Center, Hunstville, Ala. October 23. Online. Available at: http://www.nasa.msfc.nasa.gov/news

Stern, E. (1990) *The U-137 Incident: A Study in Swedish Crisis Management*. Stockholm: International Graduate School.

Stern, E. (1997) "Crisis and Learning: A Conceptual Balance Sheet." *Journal of Contingencies and Crisis Management*, 5(2):70.

Stern, E. (1999) *Crisis Decisionmaking: A Cognitive Institutional Approach. Stockholm Studies in Politics 66*. Diss. Department of Political Science, Stockholm University.

Stern, E., and F. Bynander, eds. (1998) *Crisis and Internationalisation: Eight Crises Studied From a Cognitive-Institutional Perspective*. Stockholm: ÖCB [Swedish Agency for Civil Emergency Planning].

Stern, E., and D. Hansén, eds. (2000) *Crisis Management in a Transitional Society: The Latvian Experience*. Stockholm: National Defense College of Sweden.

Stern, E. and Nohrstedt, D., eds. (1999) *Crisis Management in Estonia: Theory and Practice*. Stockholm: ÖCB [The Swedish Agency for Civil Emergency Planning].

Stern, E., and B. Sundelius. (1992) "Managing Asymmetrical Crisis: Sweden, the USSR, and U-137." *International Studies Quarterly*. 36:213–239.

Stern, E., and B. Sundelius. (1994) "The Essence of Groupthink." *Mershon International Studies Review* 38:101–107.

Stern, E., and B. Sundelius. (1997) "Sweden's Twin Monetary Crises of 1992: Rigidity and Learning in Crisis Decision Making." *Journal of Contingencies and Crisis Management*. 5(1):32–48.

Stern, E., and B. Sundelius. (2002) "Crisis Management Europe: An Integrated Regional Research and Training Program." *International Studies Perspectives* 3: 71–88.

Stern, E., and B. Verbeek, eds. (1998) "Whither the Study of Governmental Politics in Foreign Policymaking: A Symposium." *Mershon International Studies Review* 42:205–55.

Stewart, P. C., C. F. Hermann, and M. G. Hermann. (1989) "Modelling the 1973 Soviet Decision to Support Egypt." *American Political Science Review* 83(1):35–59.

Stockholms Stads Utrednings och Statistikkontoret (USK) [City of Stockholm's Investigational and Statistical Office]. (2000a). "Befolkning 1985–1998" [Population 1985–1998]. Online. Available at: http://www.usk.stockholm.se/internet/befolk/nytab13.htm

Stockholms Stads Utrednings och Statistikkontoret (USK) [City of Stockholm's Investigational and Statistical Office]. (2000b). "Befolkningsprognos 1999–2007 för Stockholm" [The 1999–2007 Population Prognosis for Stockholm]. Online. Available at: http://www.usk.stockholm.se/internet/publ/stat_utg/befpr99.htm

Sugrue, T. (1996) *The Origins of the Urban Crisis: Race and Inequality in Postwar Detroit*. Princeton, N.J.: Princeton University Press.

Sullivan, A. (1997) "Maori Politics and Government Policies." In *New Zealand Politics in Transition*. Ed. Raymond Miller. Auckland: Oxford University Press.

Sundelius, B., E. Stern, and F. Bynander. (1997) *Krishantering på Svenska: Teori och Praktik* [*Crisis Management the Swedish Way: Theory and Practice*]. Stockholm: Nerenius & Santérus.

Sustainable urban development (URBAN) A5-0026/2000. (2000) European Parliament resolution on the draft communication from the Commission to the Member States laying down guidelines for a Community Initiative concerning economic and social regeneration of cities and of neighborhoods in crisis in order to promote sustainable urban development (URBAN) (COM(1999) 477-C5-0242/1999—1999/2177(COS)). *Official Journal-European Communities Information and Notices C*, 43(339):44–46.

Svedin, L. (1998) "Mc-krig i Sverige och Norden 1994–97" [Motorcycle Gang Wars in Sweden and the Nordic Countries 1994–97] In *Crisis and Internationalisation: Eight Crises Studied From a Cognitive-Institutional Perspective*. Ed. E. Stern and F. Bynander. Stockholm: ÖCB [Swedish Agency for Civil Emergency Planning].

Svedin, L. (1999) "Crisis Mitigation and Community Resilience: The Red River Flood of '97." Unpublished report. Crisis Management Europe: National Crisis Management from an International Perspective.

Sverige Direkt. (2000) "Kommuner." Online. Available at: http://www.sverigedirekt.riksdagen.se/lanksamling/kommuner.asp

Swaffield, S. (1998) "Contextual Meanings in Policy Discourse: A Case Study of Language Use in the New Zealand High Country." *Policy Sciences* 31:199–224.

Sylvan, D., and J. Voss, eds. (1998) *Problem Representation in Foreign Policy Decision Making*. Cambridge, U.K.: Cambridge University Press.

Tetlock, P. (1985) "Accountability: The Neglected Social Context of Judgement and Choice." In *Research in Organisational Behavior*. Ed. L. Cummings, and B. Staw, 7:297–332.

Tetlock, P. (1992) "The Impact of Accountability on Judgement and Choice: Toward a Social Contingency Model." *Advances in Experimental Social Psychology* 25. New York: Academic Press.

Tetlock, P., and A. Belkin, eds. (1986) *Counterfactual Thought Experiments in Word Politics: Logical, Methodological, and Psychological Perspectives*. Princeton, N.J.: Princeton University Press.

Theutenberg, B. (1982) *Polarområdena: resursjakt och folkrätt*. Stockholm: Utrikespolitiska institutet.

Theutenberg, B. (1986) *Folkrätt och säkerhetspolitik*. Stockholm: Norstedt i samarbete med Kungl. Krigsvetenskapsakad.

Tizard, J. (1998) "Parliamentary Session of the Government of New Zealand." Hansard (02/24/98). Online. Available at: http//rangi.knowledge-basket.co.nz/hansard/han/text/1998/02/24-020.html

Transpower. (1997) "Customers to Determine Electricity Security from National Grid." News, Sept. 4, 1997. Available at: http://www.transpower.co.nz/news/media/199712_4Customers_to.shtml

Transpower. (1999) "Statement of Corporate Intent" Available at: http://www.transpower.co.nz/publications/sci.shtml

Turner, B. (1978) *Man-Made Disasters*. London: Wykeham.

Turner, B., and N. Pidgeon. (1997) *Man-Made Disasters* (2nd ed.). Oxford, U.K.: Buckworth-Heinemann.

Uddin, M. N., and E. M. Jones. (2000) "Sanitation: the Crisis of the Urban Poor." Wedc Conference, 26:299–302.

Ullberg, S. (1998) "Limakrisen, 1997" [The Lima Crisis]. In *Crisis and Internationalisation: Eight Crises Studied From a Cognitive-Institutional Perspective*. Ed. Eric Stern and Fredrik Bynander. Stockholm: ÖCB [The Swedish Agency for Civil Emergency Planning].

United Nations Development Program (UNDP). (1999) *Human Development Report 1999*. New York: Oxford University Press

Underwood M. M., and J. Kalafat. (2002) "Crisis Intervention in a New Context: New Jersey Post–September 11, 2001." *Brief Treatment and Crisis Intervention* 2(1): 75–84(10).

United States Congress. (1992) "Rhetoric, Not Radical Change, Follows L.A. Riots." *Congressional Quarterly Weekly Report*, 50(19):1247.

Vandenbroucke, L. (1984) "Anatomy of a Failure: the Decision to Land at the Bay of Pigs." *Political Science Quarterly* 99(3):471–91.

Vandenbroucke, L. (1993) *Perilous Options*. Oxford, U.K.. Oxford University Press.

Vanerburg, W. (1988) "In Richard Day." In *Democratic Theory and Technological Society.* Ed. Ronald Beiner and Joseph Masciulli. Armonk, N.Y.: M. E. Sharpe.

Venter, N. (1998) "ECNZ Split Leaves Future Sales Open." *The Dominion*. April 8. Wellington: Newspapers Limited.

Vertzberger, Y. (1990) *The World In Their Minds: Information Processing, Cognition, and Perception in Foreign Policy Decisionmaking*. Stanford, Calif.: Stanford University Press.

Wallis, A. (1998) "The Heat." *New Zealand Engineering*. April: 6–8.

Warby, R. (1998) *Operation Energy Report*. Auckland: New Zealand Fire Service.

Warren, R. (2001) "Dealing with Densities: Automated Metrolinks and Urban Oases." *Journal of Urban Technology*, 1 April, 8(1):23–47.

Waters, J. A. (2002) "Moving Forward From September 11: A Stress/Crisis/Trauma Response Model." *Brief Treatment and Crisis Intervention*, March, 2(1):55–74(20).

Waugh, W. L. (1988) "Current Policy and Implementation in Disaster Preparedness." In *Managing Disaster: Strategies and Policy Perspectives*. Ed. L. K. Comfort, 111–125. Durham, N.C.: Duke University Press.

Widmayer, P. (2000) "Transforming A Global City for the Information Society: Metropolitan Chicago at the Crossroads." *Journal of Urban Technology*, 1 August, 7(2):23–42.

Wildavsky, A. (1988) *Searching for Safety*. New Brunswick, N.J.: Transaction Books.

Wilkie, D. (1997) "Preface to the 'Stage One Report.'" *Auckland Engineering Lifelines Project*, July 1997. Auckland: Auckland Regional Council.

Winston, P. (1998) "Parliamentary Session of the Government of New Zealand." Hansard (02/24/98). Online. Available at: http//rangi.knowledge-basket.co.nz/hansard/han/text/1998/02/24-020.html

The World & I. (1991) "Special Report: Solving America's Urban Crisis," 6(6):18.

Zimmerman, R. (2001) "Social Implications of Infrastructure Network Interactions." *Journal of Urban Technology*, 1 December, 8(3):97–119.

PERSONAL INTERVIEWS

Bengtsson, Gunnar 12/12/96	Former Director, Swedish Radiological Protection Insitute
Brown, Tracy 11/17/98	Area Court Manager of the District Court Department for Courts, Auckland
Cheng, Raymond 11/16/98	Manager, Network Customer Service, Mercury Energy Limited, Auckland
de Abaitua, Paul 11/16/98	Manager, CBD Supply Diversity Project, Mercury Energy Limited, Auckland
Gardiner, James 11/18/98	Chief Reporter, *The New Zealand Herald*, Auckland

Heanan, Peter 11/20/98	Director of the Australia Division, New Zealand Ministry of Foreign Affairs, Wellington
Heavey, Linda 11/18/98	Promotions Co-ordinator, Karangahape Road (K' Road) Business Association, Auckland
Henare, Fred 11/16/98	Assistant Commissioner, Civil Defence, Auckland Division
Hope, Virginia 11/18/98	Programme Manager, Environmental Health and Medical Officer of Health, Public Health Physician, Auckland Healthcare Services Ltd., Auckland
Hutchinson, Evan 11/18/98	General Manager, Regional Parks, Auckland Regional Council
Lindsay, John 11/17/98	Emergency Management Team Leader, Auckland City Council
May, John 11/18/98	Manger, Public Relations on the Auckland City Council, and Mayor Les Mills' PR person during the Auckland Blackout
Manley, Clive 11/17/98	Manager, Democracy Services, Auckland City Council
McKeen, Graham 11/18/98	Auckland Regional Council
Moore, Dallas 11/20/98	Manager, Planning and Public Education Ministry of Civil Defence Wellington
Moore, John 11/16/98	Operations Manager, Network Customer Services Mercury Energy Limited, Auckland
Neller, Marion 11/17/98	Court Manager, High Court Department for Courts, Auckland
Norton, John 11/20/98	Director for Management and Civil Defence, Wellington
Pritchards, Kathy 11/18/98	Food Programme Manager, Auckland Public Health and Healthcare Services Ltd., Auckland
Rangihika, Rerangi 11/18/98	Police Chief Superintendent, Auckland Police Department
Small, Rob 11/18/98	Auckland Regional Council, Auckland General Manager of the Regional Parks
Warby, Ray 11/17/98	Chief Fire Officer, Auckland in City East New Zealand Fire Services
Wilson, Lloyd 11/16/98	Operations Engineer, Network Customer Service, Mercury Energy Limited, Auckland
Wood, Ian 11/18/98	Hazards Management, Auckland Regional Council, Auckland

Index

Auckland Unplugged